POLYMER SCIENCE ANI

THEORETICAL ESTIMATION OF ACIDIC FORCE OF LINEAR OLEFINS OF CATIONIC POLYMERIZATION

POLYMER SCIENCE AND TECHNOLOGY

Additional books in this series can be found on Nova's website under the Series tab.

Additional E-books in this series can be found on Nova's website under the E-books tab.

BIOCHEMISTRY RESEARCH TRENDS

Additional books in this series can be found on Nova's website under the Series tab.

Additional E-books in this series can be found on Nova's website under the E-books tab.

POLYMER SCIENCE AND TECHNOLOGY

THEORETICAL ESTIMATION OF ACIDIC FORCE OF LINEAR OLEFINS OF CATIONIC POLYMERIZATION

V. A. BABKIN
V. U. DMITRIEV
D. S. ANDREEV
V. T. FOMICHEV
E. S. TITOVA
AND
G.E. ZAIKOV

Nova Science Publishers, Inc.
New York

For permission to use material from this book please contact us:
Telephone 631-231-7269; Fax 631-231-8175
Web Site: http://www.novapublishers.com

LIBRARY OF CONGRESS CATALOGING-IN-PUBLICATION DATA

Theoretical estimation of acidic force of linear olefins of cationic
polymerization / [edited by] G.E. Zaikov, Babkin Vladimir Alexandrovich.
p. cm.
Includes index.
ISBN 978-1-61209-578-3 (softcover)
1. Alkenes. 2. Addition polymerization. 3. Quantum chemistry. I.
Zaikov, G. E. (Gennadii Efremovich), 1935- II. Babkin, V. A. (Vladimir
Aleksandrovich), 1952-
QD305.H7T43 2011
547'.412--dc22

2011003802

Published by Nova Science Publishers, Inc. † New York

CONTENTS

PREFACE

This new book presents research results on the quantum-chemical calculation of linear olefins of cationic polymerization (a geometrical and electronic structure, the general and electronic energy, distribution of charges to atoms and so forth). In addition, acid force of these monomers is theorized. The book is intended for science officers, competitors, post-graduate students and persona working for a doctorate degree in the field of polymer chemistry.

ABSTRACT

In monograph results of quantum-chemical calculation of linear olefins of cationic polymerization (a geometrical and electronic structure, the general and electronic energy, distribution of charges to atoms and so forth) are presented. Acid force of these monomers is theorized.

The Monograph is intended for science officers, competitors, post-graduate students and person working for doctor's degree, working in the field of polymer chemistry.

INTRODUCTION

The Syntheses of polymers (polyethylene, polypropylene and other) from linear monomers is a complex catalytic process of cationic polymerization of olefins [1 - 3]. It consists of 3 stages: initiation, growing and breakaway of circuit. The necessary components of these reactions are: catalyst, cocatalyst, solvent and monomer. The Influence of acid power of the initiator (addition compound catalyst + promoter) on efficiency of the process of cationic polymerization of linear olefins is studied [1-3]. The influence of acid power of the monomer on process of cationic polymerization of linear olefins and estimation their acid power not studied.

In this connection the aim of this monograph is quantum-chemical calculation and estimation acid power of linear monomers of cationic polymerization by the most comprehensible methods of quantum chemistry.

Chapter 1

QUANTUM-CHEMICAL METHODS OF CALCULATION

Quantum-chemical methods of calculation allow defining not only a relative arrangement of atoms in geometrical structure of a molecule, but also a lot of parameters of investigated connections influencing behaviour in various chemical reactions. One or several basic sizes (coordinate of reaction, electronic density on atoms of a molecule, power characteristics, the dipole moment) is possible to evolve from this row of parameters and to lead ordering results of quantum-chemical calculations in comparison with physicochemical characteristics of investigated compounds.

The Most propagated theoretical quantum-chemical methods studying of molecular systems are methods of molecular orbitals (MO) in approach LCAO (a linear combination of atomic orbitals). Methods of approach LCAO subdivide all on two greater groups: strict nonempirical and semiempirical methods of SCF (the self-consistent field).

Values only fundamental physical constants are required for nonempirical calculations. Nonempirical calculations are necessary now for three overall objectives: first, for finding-out of conformity between the positioned experimental facts and the theory; secondly, for calculation of properties of molecular structures or effects which it is inconvenient or it is impossible to measure experimentally (transitive conditions of reaction); thirdly, for a substantiation of development of semiempirical schemes of calculation. The two first purposes are characteristic and for semiempirical methods of SCF in which experimental data are used for an estimation of segregate integrals. The following approaches are supposed:

1) Some groups of electrons are not considered by obvious image. For

example, in calculation consider only valent electrons (valent approach);

2) Some integrals or are not considered (are accepted equal to zero), or are expressed through empirical parameters.

These Assumptions and replacements cannot be any. The basic criterion of approach should be an invariance of calculation at rotation of coordinate axes and movings (translations) of molecular systems from one point of space in another without change of structure.

In this connection a series of levels of approaches is possible [4].

1) Approaches, leading that results of calculations become not invariant as concerning rotation of coordinate axes, and hybridization of atomic orbitals (AO).
2) Approaches, leading preservation of invariancy of results concerning rotation of coordinate axes, but to infringement of invariancy concerning AO.
3) Approaches, conserving invariancy of results rather and rotations of coordinate axes, and hybridizations of AO.
4) Approaches, conserving invariancy of results at any orthogonal transformations of basis of AO, including at transition to orbitals symmetry.

In practice preservation of invariancy of the second and third levels is sufficient.

All semiempirical methods of calculation can be divided into methods of a full differential overlapping and a zero differential overlapping (ZDO). The Most propagated in the first group is expanded Heckle method (EHT) [4, 5.] An Electron-electron interaction is not considered in this method. The Quantity of settlement integrals of interelectronic interaction essentially contracts in approach ZDO comparison with nonempirical methods.

Known variants of ZDO basically keep within Popl classification. Three levels of approaches, conserving for quite correct practical use sufficient invariancy of decisions it is possible to evolve with this classification [4]. The Idea is based on supervision, that in all nonempirical calculations two-electron molecular integrals of a kind *(ab / cd)* [5] in which for AO conditions $a = b$ and $c = d$ are satisfied, always many times more than those integrals in which even one of two conditions is not carried out and consists in the offer to reject the last. At rejection of all without exception of conditions the method has received the name of complete neglect of differential overbreaking (CNDO). At preservation of those conditions in which orbitals *a* and *b* (or *c* and *d*) though are various, but belong to the same atom - the method has received the name of neglect of diatomic differential overlapping (NDDO). If only a part from conditions to conserve the method has received the name of a method of partial neglect of differential overlapping (INDO) [5].

Methods CNDO and INDO badly reproduce heats of atomization and orbital energys. Dewar has come to conclusion about necessity reparametrization method INDO with the purpose of reception of reliable values of heats of formation and molecular geometry. It promoted development of method MINDO. Parameters of method MINDO differ from scheme INDO introduction of some empirical dependences for integrals and energy of pushing away of skeletons. They get out so that to achieve the best consent of the calculated and experimental values of heats of formation and geometrical characteristics of a wide class of standard compounds. Various parameterizations of method MINDO exist. The Greatest distribution was received methods MNDO and MINDO/3. Method MINDO/3 is the best semiempirical a method for construction of surfaces of a potential energy and studying of mechanisms of reactions.

Quantum-chemical method MNDO has well proved in calculations nonlimiting compounds and molecules containing unshared pairs on the next atoms. It yields more exact values of valent corners and correct sequence of levels molecular orbitals. MNDO correctly reproduces relative stability of the compounds containing double bonds. It is very essential at studying elementary certificates of cationic polymerization of olefins. Polymerization represents repeated transformations π -bonds of monomer in σ-bonds.

But MINDO/3 and MNDO use empirical function for nuclear-nuclear pushing away which is absolutely homogeneous on greater distances. To correct big enough pushing away of kernels on distances Van der Waals type in method AM1 are added a little of Gaussians (which decay much more quickly than initial function). Characteristics for molecules including hydrogen bonds considerably improve.

New method PM3 represents global reparametrization all elements in processing AM1 (in difference from partially divided optimization till now used in methods AM1 and MNDO) [6].

The Opportunity of a choice of various atomic orbitals (basic sets) is stipulated in nonempirical methods in difference from semiempirical programs ZDO [7]. All modern nonempirical programs contain basic sets of orbitals of Gauss type (OGT). Basic sets to which each AO is presented by several Gauss functions of distribution of electronic density, have significant advantages before basic sets (for example, equivalent Slater orbitals) at calculation of one-and two-centered integrals in detail described in [8, 10] and allow to save machine time.

Population of areas of an overlapping, charges on atoms and so on often appear inexact at carrying out of nonempirical calculations in greater basic sets. The analysis of population by Milliken represents a procedure of division of electron density between atoms and atomic orbitals. This procedure is based on values of factors AO in molecular orbitals and approach of uniform distribution of density of an overlapping (Milliken approach). The Procedure can lead to physically unreasonable values of atomic charges at inclusion in a basic set of diffuse functions. This complication is connected with the big extent of diffuse orbitals, next atoms covering space. In essence such orbitals improve the description of the next atoms and substantially concern to all molecule, notwithstanding what they are aligned on one of atoms. All electronic density connected by diffuse orbital, is attributed to atom on which this orbital is aligned at carrying out of the Milliken analysis of population. The yielded difficulty can be bypassed, having divided all space on area and having attributed their segregate atoms. The full electronic density in the field of, carried to the yielded atom, can be integrated for calculation of a charge on atom. Value of a charge depends on a way of division of electronic density between atoms. The problem of ambiguity of division of space on atomic areas has very specific character, and standard and its formalized decision is not found yet. "Loan" of free orbitals at the next atoms can lead to mistakes in an estimation of energy of bonds. Such mistakes name superimposed mistakes of a basic set. They arise most often at linkage of the atoms strongly differing on electronegativity, for example at linkage of lithium and fluorine.

Two above described methods of quantum-chemical calculations MNDO (as the fastest now) and AB INITIO in base 6-311G ** (as the basis in which is marked the best correlation between the maximal charge on atom of hydrogen q_{max}^{H+} and pKa - a universal parameter of acidity, the employee for a theoretical estimation of acid force of studied monomers) us have been chosen for research of a geometrical and electronic structure of linear olefins of cationic polymerize-tion at the analysis [9].

Chapter 2

QUANTUM-CHEMICAL CALCULATION OF LINEAR OLEFINS OF CATIONIC POLYMERIZATION

2.1. CALCULATION BY METHOD AB INITIO

The Aim of this chapter is quantum-chemical calculation of the row linear olefins: ethylene, propylene, butene−1, butene-2, pentene-1, geksene-1, geptene-1, oktene-1, nonene-1, dekene-1, theoretical estimation their acid power, study of its influence upon efficiency of cationic polymerization and quality got polymer [11].

Quantum-chemical calculation of linear monomer of cationic polymerization was made by method AB INITIO in base 6-311G**[12] with optimization of the geometries by all parameters by standard gradient method in approach the insulated molecule in gas phase. The expression pKa = 49,4-134,61·q_{max}^{H+} [6] (where pKa - an universal factor to acidity, $q_{H^+}^{max}$ - a maximum positive charge on atom of the hydrogen) used for the estimation of acid power of linear olefins. The base 6-311G** gives the best factor of correlations R=0.97. Program MacMolPlt was used for visual presentation of the model of the molecule [14].

RESULTS AND DISCUSSION

Geometric and electronic structure of linear olefins: ethylene, propylene, butene−1, butene-2, pentene-1, geksene-1, geptene-1, oktene-1, nonene-1, dekene-1, their quantum-chemical parameters (E_0- a general energy of the molecular system, $E_{of\ bond}$ – a total energy of all bonds of olefin, D- a dipole moment, q_H- charges on α- and β- carbon atom of olefin, q_{max}^{H+}- a maximum positive charge on atom of the hydrogen of the monomer) and pKa – an universal factor to acidity- presented in tables 2.1.1-2.1.8 and on drawings 2.1.1-2.1.7.

Table 2.1.1. Optimized lengths of bonds, valency corners, charges of atoms of the molecule of ethylene

Length of bond	R, A	Valency corners	Grad	Atom	Charge (by Milliken)
C(1)-C(2)	1,31			1 C	-0.22
C(1)-H(1)	1,07	H(1)C(1)C(2)	122	2 C	-0.22
C(1)-H(2)	1,07	H(2)C(1)C(2)	122	3 H	0.11
C(2)-H(3)	1,07	H(3)C(2)C(1)	122	4 H	0.11
C(2)-H(4)	1,07	H(4)C(2)C(1)	122	5 H	0.11
				6 H	**0.11**

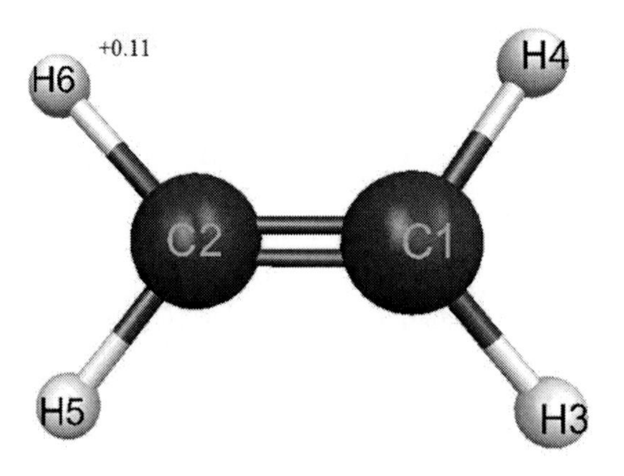

Figure 2.1.1. Geometric and electronic structure of molecule of ethylene.(E_0 = -204558 kDg/mol, E_{el}= -292855 kDg/mol).

Table 2.1.2. Optimized lengths of bonds, valency corners, charges of atoms of the molecule of propylen

Length of bond	R, A	Valency corners	Grad	Atom	Charge (by Milliken)
C(1)-C(2)	1,31			1 C	-0.21
C(2)-C(3)	1,50	C(3)C(2)C(1)	125	2 C	-0.17
C(1)-H(4)	1,07	H(4)C(1)C(2)	121	3 C	-0.21
C(1)-H(5)	1,07	H(5)C(1)C(2)	121	4 H	0.10
C(2)-H(6)	1,07	H(6)C(2)C(1)	118	5 H	0.10
C(3)-H(7)	1,08	H(7)C(3)C(2)	110	6 H	0.10
C(3)-H(8)	1,08	H(8)C(3)C(2)	111	7 H	0.10
C(3)-H(9)	1,08	H(9)C(3)C(2)	110	8 H	0.10
				9 H	**0.10**

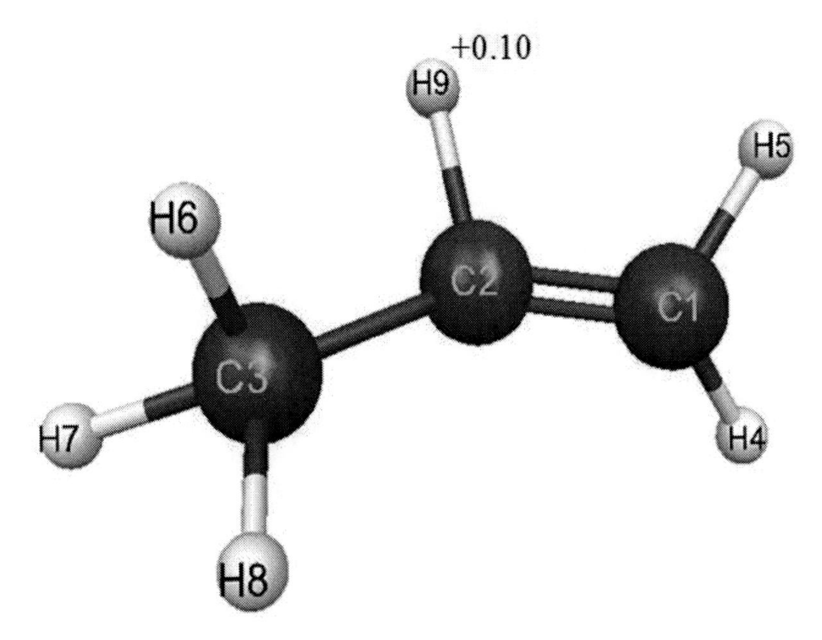

Figure 2.1.2. Geometric and electronic structure of the molecule of propylene (E_0= -306903 kDg/mol, E_{el}= -493272 kDg/mol).

Table 2.1.3.Optimized lengths of bonds, valency corners, charges of atoms of the molecule of butene–1

Length of bond	R, A	Valency corners	Grad	Atom	Charge (by Milliken)
C(1)-C(2)	1,32			1 C	-0.20
C(2)-C(3)	1,50	C(3)C(2)C(1)	125	2 C	-0.16
C(3)-C(4)	1,53	C(4)C(3)C(2)	112	3 C	-0.16
C(1)-H(5)	1,07	H(5)C(1)C(2)	121	4 C	-0.24
C(1)-H(6)	1,08	H(6)C(1)C(2)	121	5 H	0.10
C(2)-H(7)	1,08	H(7)C(2)C(1)	118	6 H	0.10
C(3)-H(8)	1,08	H(8)C(3)C(2)	109	**7 H**	**0.10**
C(3)-C(9)	1,08	H(9)C(3)C(2)	108	8 H	0.10
C(4)-H(10)	1,08	H(10)C(4)C(3)	110	9 H	0.10
C(4)-H(11)	1,08	H(11)C(4)C(3)	111	10 H	0.09
C(4)-H(12)	1,08	H(12)C(4)C(3)	110	11 H	0.09
				12 H	0.09

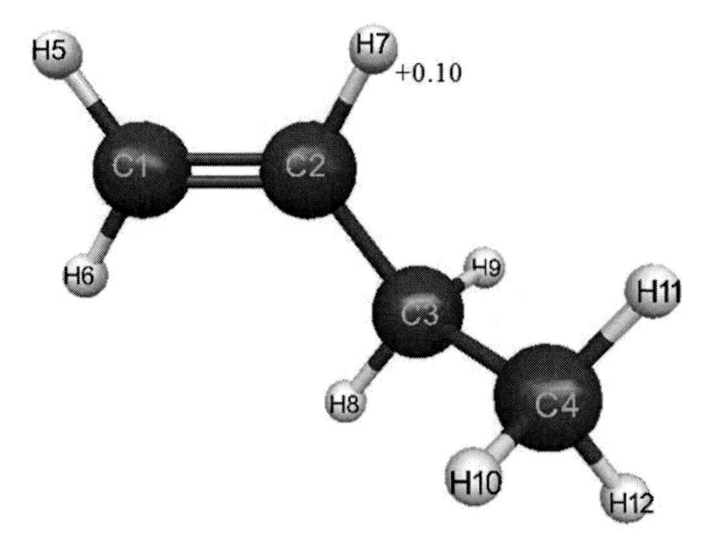

Figure 2.1.3. Geometric and electronic structure of the molecule of butene–1 (E_0= =-409247,289 kDg/mol, E_{el}= -717329,382 kDg/mol).

Table 2.1.4. Optimized lengths of bonds, valency corners, charges of atoms of the molecule of butene–2

Length of bond	R, A	Valency corners	Grad	Atom	Charge (by Milliken)
C(1)-C(2)	1,50			1 C	-0.20
C(2)-C(3)	1,31	C(3)C(2)C(1)	125	2 C	-0.16
C(3)-C(4)	1,50	C(4)C(3)C(2)	125	3 C	-0.16
C(1)-H(5)	1,08	H(5)C(1)C(2)	111	4 C	-0.20
C(1)-H(6)	1,08	H(6)C(1)C(2)	110	5 H	0.09
C(1)-H(7)	1,08	H(7)C(1)C(2)	110	6 H	0.09
C(2)-H(8)	1,08	H(8)C(2)C(3)	118	7 H	0.09
C(3)-C(9)	1,08	H(9)C(3)C(2)	118	**8 H**	**0.09**
C(4)-H(10)	1,08	H(10)C(4)C(3)	110	9 H	0.09
C(4)-H(11)	1,08	H(11)C(4)C(3)	110	10 H	0.09
C(4)-H(12)	1,08	H(12)C(4)C(3)	111	11 H	0.09
				12 H	0.09

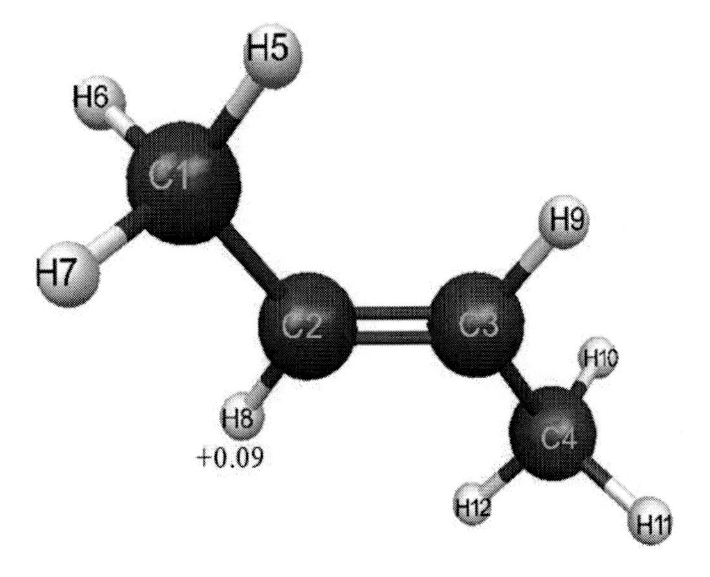

Figure 2.1.4. Geometric and electronic structure of the molecule of butene–2 (E0= -409247 kDg/mol, Eel = -713922 kDg/mol).

Table 2.1.5. Optimized lengths of bonds, valency corners, charges of atoms of the molecule of pentene -1

Length of bond	R, A	Valency corners	Grad	Atom	Charge (by Milliken)
C(1)-C(2)	1,31			1 C	-0.19
C(2)-C(3)	1,50	C(3)C(2)C(1)	125	2 C	-0.15
C(3)-C(4)	1,53	C(4)C(3)C(2)	112	3 C	-0.15
C(4)-C(5)	1,52	C(5)C(4)C(3)	112	4 C	-0.20
C(1)-H(6)	1,07	H(6)C(1)C(2)	121	5 C	-0.24
C(1)-H(7)	1,07	H(7)C(1)C(2)	121	6 H	0.10
C(2)-H(8)	1,08	H(8)C(2)C(3)	115	7 H	0.10
C(3)-H(9)	1,08	H(9)C(3)C(2)	109	**8 H**	**0.10**
C(3)-C(10)	1,08	H(10)C(3)C(2)	108	9 H	0.09
C(4)-H(11)	1,08	H(11)C(4)C(3)	109	10 H	0.10
C(4)-H(12)	1,08	H(12)C(4)C(3)	108	11 H	0.09
C(5)-H(13)	1,08	H(13)C(5)C(4)	111	12 H	0.09
C(5)-H(14)	1,08	H(14)C(5)C(4)	111	13 H	0.09
C(5)-H(15)	1,08	H(15)C(5)C(4)	111	14 H	0.09
				15 H	0.09

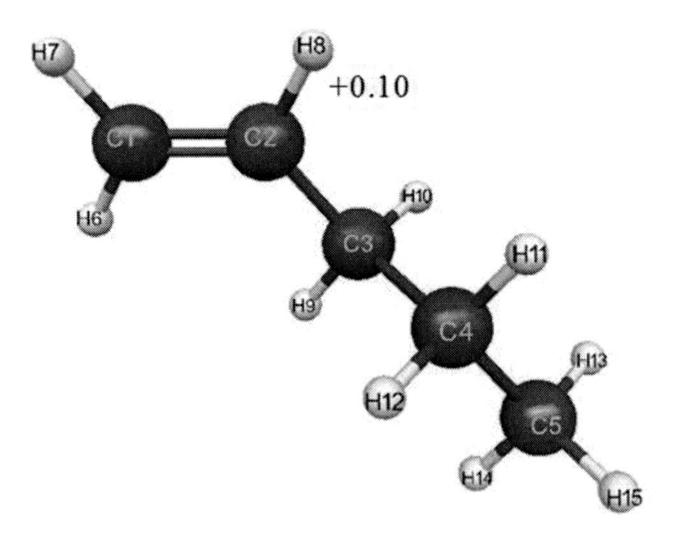

Figure 2.1.5. Geometric and electronic structure of the molecule of pentene -1 (E0= -511566 kDg/mol, Eel = -958160 kDg/mol).

Table 2.1.6. Optimized lengths of bonds, valency corners, charges of atoms of the molecule of geksene-1

Length of bond	R, A	Valency corners	Grad	Atom	Charge (by Milliken)
C(1)-C(2)	1,31			1 C	-0.20
C(2)-C(3)	1,51	C(3)C(2)C(1)	128	2 C	-0.16
C(3)-C(4)	1,53	C(4)C(3)C(2)	117	3 C	-0.13
C(4)-C(5)	1,53	C(5)C(4)C(3)	115	4 C	-0.22
C(5)-C(6)	1,52	C(6)C(5)C(4)	114	5 C	-0.19
C(1)-H(7)	1,07	H(7)C(1)C(2)	122	6 C	-0.24
C(1)-H(8)	1,07	H(8)C(1)C(2)	120	7 H	0.10
C(2)-H(9)	1,08	H(9)C(2)C(1)	114	8 H	0.10
C(3)-H(10)	1,08	H(10)C(3)C(2)	107	9 H	0.10
C(3)-H(11)	1,08	H(11)C(3)C(2)	108	**10 H**	**0.11**
C(4)-H(12)	1,08	H(12)C(4)C(3)	109	11 H	0.10
C(4)-H(13)	1,08	H(13)C(4)C(3)	107	12 H	0.10
C(5)-H(14)	1,08	H(14)C(5)C(4)	110	13 H	0.10
C(5)-H(15)	1,08	H(15)C(5)C(4)	107	14 H	0.10
C(6)-H(16)	1,08	H(16)C(6)C(5)	112	15 H	0.09
C(6)-H(17)	1,08	H(17)C(6)C(5)	111	16 H	0.09
C(6)-H(18)	1,08	H(18)C(6)C(5)	110	17 H	0.08
				18 H	0.09

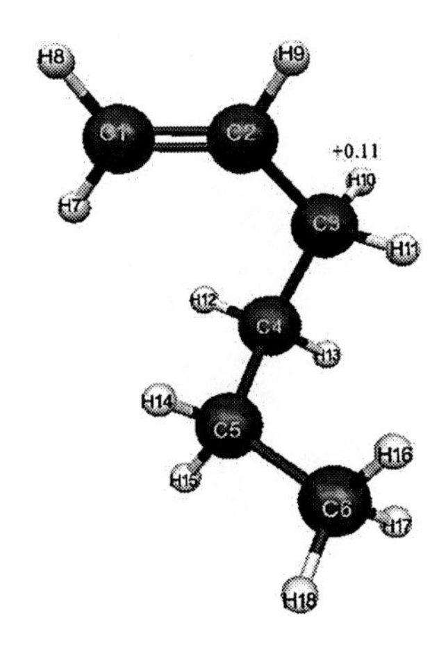

Figure 2.1.6. Geometric and electronic structure of the molecule of geksene-1 (E0= -613884 kDg/mol, Eel = -1236259 kDg/mol).

Table 2.1.7. Optimized lengths of bonds, valency corners, charges of atoms of the molecule of geptene-1

Length of bond	R, A	Valency corners	Grad	Atom	Charge (by Milliken)
C(1)-C(2)	1,31			1 C	-0.20
C(2)-C(3)	1,51	C(3)C(2)C(1)	128	2 C	-0.16
C(3)-C(4)	1,53	C(4)C(3)C(2)	117	3 C	-0.13
C(4)-C(5)	1,53	C(5)C(4)C(3)	116	4 C	-0.23
C(5)-C(6)	1,53	C(6)C(5)C(4)	116	5 C	-0.18
C(6)-C(7)	1,52	C(7)C(6)C(5)	114	6 C	-0.20
C(1)-H(8)	1,07	H(8)C(1)C(2)	122	7 C	-0.24
C(1)-H(9)	1,07	H(9)C(1)C(2)	120	8 H	0.10
C(2)-H(10)	1,08	H(10)C(2)C(3)	114	9 H	0.10
C(3)-H(11)	1,08	H(11)C(3)C(2)	107	10 H	0.09
C(3)-H(12)	1,08	H(12)C(3)C(2)	108	**11 H**	**0.11**
C(4)-H(13)	1,08	H(13)C(4)C(3)	109	12 H	0.10
C(4)-H(14)	1,08	H(14)C(4)C(3)	107	13 H	0.10
C(5)-H(15)	1,08	H(15)C(5)C(4)	109	14 H	0.10
C(5)-H(16)	1,08	H(16)C(5)C(4)	107	15 H	0.10
C(6)-H(17)	1,08	H(17)C(6)C(5)	110	16 H	0.09
C(6)-H(18)	1,08	H(18)C(6)C(5)	107	17 H	0.09
C(7)-H(19)	1,08	H(19)C(7)C(6)	112	18 H	0.09
C(7)-H(20)	1,08	H(20)C(7)C(6)	111	19 H	0.09
C(7)-H(21)	1,08	H(21)C(7)C(6)	110	20 H	0.08
				21 H	0.09

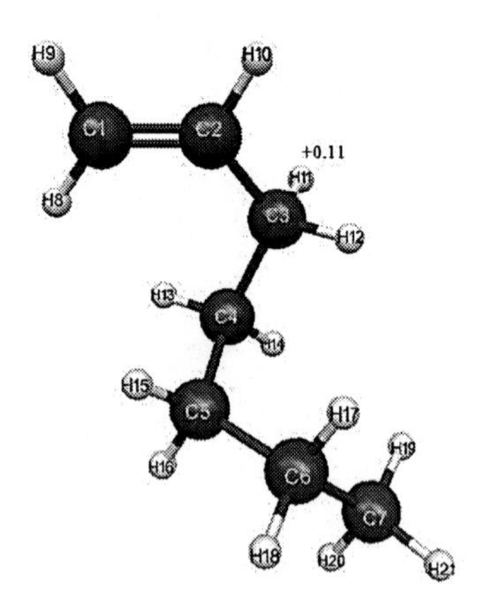

Figure 2.1.7. Geometric and electronic structure of the molecule of geptene-1 (E0= -716202 kDg/mol, Eel = -716202 kDg/mol).

Table 2.1.9. Optimized lengths of bonds, valency corners, charges of atoms of the molecule of nonene-1

Length of bond	R, A	Valency corners	Grad	Atom	Charge (by Milliken)
C(1)-C(2)	1,32			1 C	-0.20
C(2)-C(3)	1,51	C(3)C(2)C(1)	128	2 C	-0.16
C(3)-C(4)	1,53	C(4)C(3)C(2)	117	3 C	-0.13
C(4)-C(5)	1,53	C(5)C(4)C(3)	116	4 C	-0.22
C(5)-C(6)	1,53	C(6)C(5)C(4)	116	5 C	-0.18
C(6)-C(7)	1,53	C(7)C(6)C(5)	116	6 C	-0.20
C(7)-C(8)	1,53	C(8)C(7)C(6)	116	7 C	-0.19
C(8)-C(9)	1,52	C(9)C(8)C(7)	114	8 C	-0.19
C(1)-H(10)	1,08	H(10)C(1)C(2)	122	9 C	-0.24
C(1)-H(11)	1,08	H(11)C(1)C(2)	120	10 H	0.10
C(2)-H(12)	1,08	H(12)C(2)C(3)	114	11 H	0.10
C(3)-H(13)	1,08	H(13)C(3)C(2)	107	12 H	0.09
C(3)-H(14)	1,08	H(14)C(3)C(2)	108	**13 H**	**0.11**
C(4)-H(15)	1,08	H(15)C(4)C(3)	109	14 H	0.10
C(4)-H(16)	1,08	H(16)C(4)C(3)	107	15 H	0.10
C(5)-H(17)	1,08	H(17)C(5)C(4)	109	16 H	0.10
C(5)-H(18)	1,08	H(18)C(5)C(4)	107	17 H	0.10
C(6)-H(19)	1,08	H(19)C(6)C(5)	109	18 H	0.09
C(6)-H(20)	1,08	H(20)C(6)C(5)	107	19 H	0.10
C(7)-H(21)	1,08	H(21)C(7)C(6)	109	20 H	0.10
C(7)-H(22)	1,08	H(22)C(7)C(6)	107	21 H	0.09
C(8)-H(23)	1,08	H(23)C(8)C(7)	110	22 H	0.09
C(8)-H(24)	1,08	H(24)C(8)C(7)	107	23 H	0.09
C(9)-H(25)	1,08	H(25)C(9)C(8)	111	24 H	0.09
C(9)-H(26)	1,08	H(26)C(9)C(8)	110	25 H	0.08
C(9)-H(27)	1,08	H(27)C(9)C(8)	111	26 H	0.09
				27 H	0.09

Table 2.1.8. Optimized lengths of bonds, valency corners, charges of atoms of the molecule of oktene-1

Length of bond	R, A	Valency corners	Grad	Atom	Charge (by Milliken)
C(1)-C(2)	1,31			1 C	-0.19
C(2)-C(3)	1,50	C(3)C(2)C(1)	125	2 C	-0.15
C(3)-C(4)	1,53	C(4)C(3)C(2)	112	3 C	-0.15
C(4)-C(5)	1,52	C(5)C(4)C(3)	113	4 C	-0.19
C(5)-C(6)	1,52	C(6)C(5)C(4)	113	5 C	-0.18
C(6)-C(7)	1,52	C(7)C(6)C(5)	113	6 C	-0.18
C(7)-C(8)	1,52	C(8)C(7)C(6)	113	7 C	-0.19
C(1)-H(9)	1,07	H(9)C(1)C(2)	121	8 C	-0.23
C(1)-H(10)	1,07	H(10)C(1)C(2)	121	9 H	0.10
C(2)-H(11)	1,08	H(11)C(2)C(3)	109	10 H	0.10
C(3)-H(12)	1,08	H(12)C(3)C(2)	108	11 H	0.09
C(3)-H(13)	1,08	H(13)C(3)C(2)	109	12 H	0.10
C(4)-H(14)	1,08	H(14)C(4)C(3)	108	13 H	0.10
C(4)-H(15)	1,08	H(15)C(4)C(3)	109	14 H	0.09
C(5)-H(16)	1,08	H(16)C(5)C(4)	109	**15 H**	**0.10**
C(5)-H(17)	1,08	H(17)C(5)C(4)	109	16 H	0.09
C(6)-H(18)	1,08	H(18)C(6)C(5)	109	17 H	0.09
C(6)-H(19)	1,08	H(19)C(6)C(5)	109	18 H	0.09
C(7)-H(20)	1,08	H(20)C(7)C(6)	109	19 H	0.09
C(7)-H(21)	1,08	H(21)C(7)C(6)	109	20 H	0.09
C(8)-H(22)	1,08	H(22)C(8)C(7)	111	21 H	0.09
C(8)-H(23)	1,08	H(23)C(8)C(7)	111	22 H	0.08
C(8)-H(24)	1,08	H(24)C(8)C(7)	111	23 H	0.09
				24 H	0.08

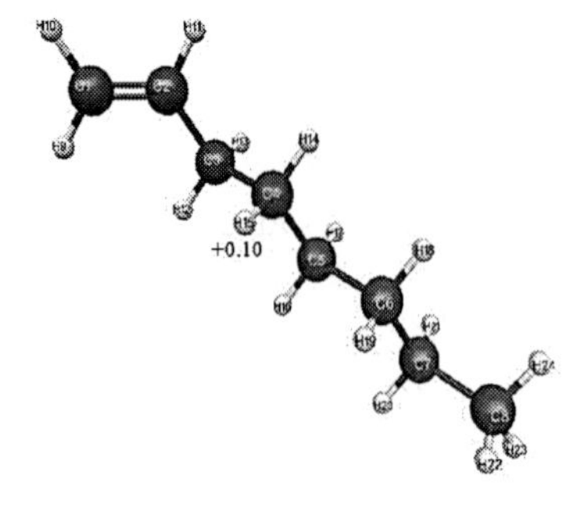

Figure 2.1.8. Geometric and electronic structure of the molecule of oktene-1 (E_0= -818547 kDg/mol, E_{el}= -1755111 kDg/mol).

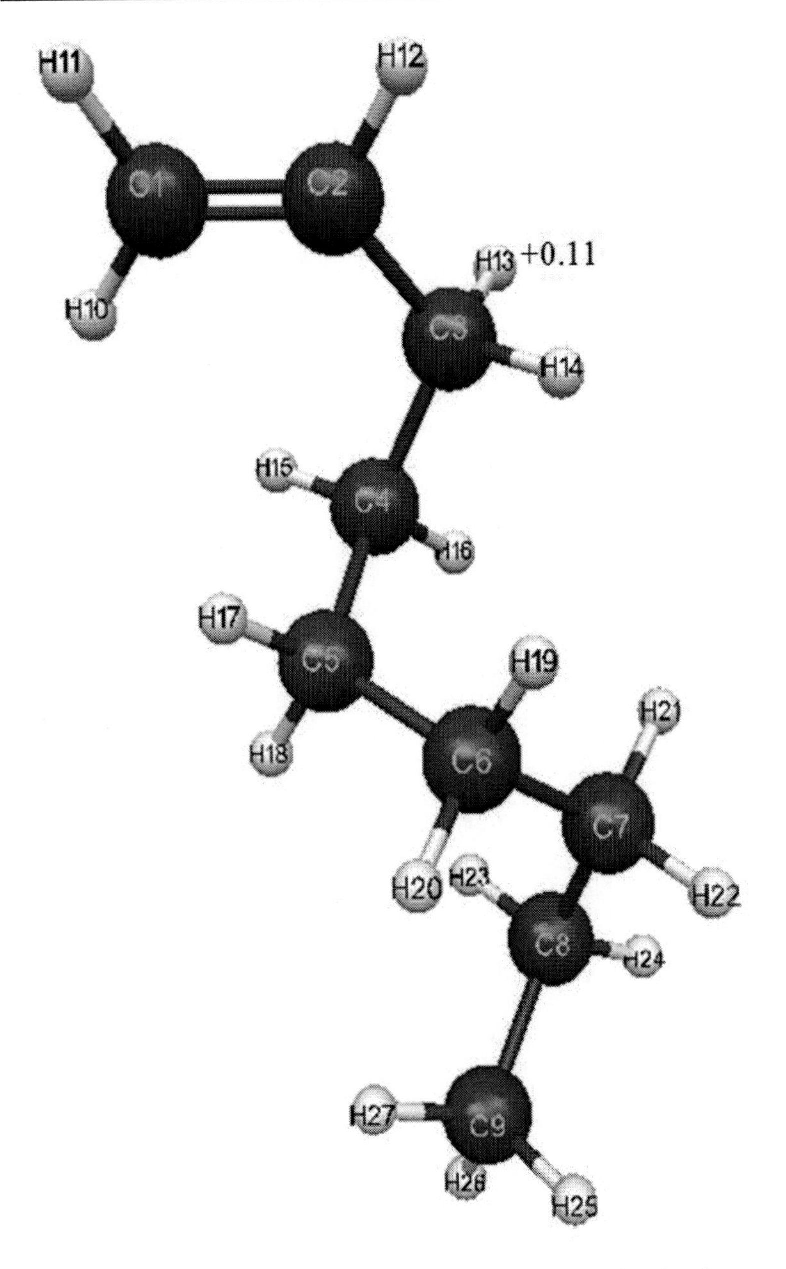

Figure 2.1.9. Geometric and electronic structure of the molecule of nonene-1 (E0= -920839 kDg/mol, Eel = -2120407 kDg/mol).

Table 2.1.10. Optimized lengths of bonds, valency corners, charges of atoms of the molecule of dekene-1

Length of bond	R, A	Valency corners	Grad	Atom	Charge (by Milliken)
C(1)-C(2)	1,32			1 C	-0.20
C(2)-C(3)	1,51	C(3)C(2)C(1)	128	2 C	-0.16
C(3)-C(4)	1,53	C(4)C(3)C(2)	117	3 C	-0.13
C(4)-C(5)	1,53	C(5)C(4)C(3)	116	4 C	-0.22
C(5)-C(6)	1,53	C(6)C(5)C(4)	116	5 C	-0.18
C(6)-C(7)	1,53	C(7)C(6)C(5)	116	6 C	-0.18
C(7)-C(8)	1,53	C(8)C(7)C(6)	114	7 C	-0.19
C(8)-C(9)	1,53	C(9)C(8)C(7)	114	8 C	-0.19
C(9)-C(10)	1,52	C(10)C(9)C(8)	114	9 C	-0.19
C(1)-H(11)	1,08	H(11)C(1)C(2)	122	10 C	-0.24
C(1)-H(12)	1,08	H(12)C(1)C(2)	120	11 H	0.10
C(2)-H(13)	1,08	H(13)C(2)C(3)	114	12 H	0.10
C(3)-H(14)	1,08	H(14)C(3)C(2)	107	13 H	0.09
C(3)-H(15)	1,08	H(15)C(3)C(2)	108	14 H	0.11
C(4)-H(16)	1,08	H(16)C(4)C(3)	109	15 H	0.10
C(4)-H(17)	1,08	H(17)C(4)C(3)	107	16 H	0.10
C(5)-H(18)	1,08	H(18)C(5)C(4)	109	17 H	0.10
C(5)-H(19)	1,08	H(19)C(5)C(4)	107	18 H	0.10
C(6)-H(20)	1,08	H(20)C(6)C(5)	109	19 H	0.10
C(6)-H(21)	1,08	H(21)C(6)C(5)	107	20 H	0.09
C(7)-H(22)	1,08	H(22)C(7)C(6)	109	21 H	0.09
C(7)-H(23)	1,08	H(23)C(7)C(6)	107	22 H	0.09
C(8)-H(24)	1,08	H(24)C(8)C(7)	110	23 H	0.09
C(8)-H(25)	1,08	H(25)C(8)C(7)	108	24 H	0.09
C(9)-H(26)	1,08	H(26)C(9)C(8)	109	25 H	0.09
C(9)-H(27)	1,08	H(27)C(9)C(8)	108	26 H	0.09
C(10)-H(28)	1,08	H(28)C(10)C(9)	111	27 H	0.09
C(10)-H(29)	1,08	H(29)C(10)C(9)	110	28 H	0.08
C(10)-H(30)	1,08	H(30)C(10)C(9)	111	29 H	0.09
				30 H	0.09

Table 2.1.11. General energy -E_0, energy of electron – E_{el}, charges on atoms - q_H, an universal factor to acidity of olefins

№п /п	Olefin	-E_0 (kDg/mol)	-E_{el} (kDg/mol)	$q_{C\alpha}$	$q_{C\beta}$	q_{max}^{H+}	pKa
1.	ethylene,	-204558	-292855	-0.22	-0.22	0,11	35
2.	propylene,	-306903	-493272	-0.21	-0.17	0.10	36
3.	butene–1,	-409247	-717329	-0,20	-0,16	0,10	36
4.	butene -2,	-409247	-713922	-0,20	-0,16	0,09	36
5.	pentene -1,	-511566	-958160	-0,19	-0,15	0,10	36
6.	geksene-1,	-613884	-1236259	-0,20	-0,16	0,11	35
7.	geptene-1,	-716202	-716202	-0,20	-0,16	0,11	35
8.	oktene-1,	-818547	-1755111	-0.19	-0.15	0.10	36
9.	nonene-1,	-920839	-2120407	-0.20	-0.16	0.10	36
10.	dekene-1	-1023184	-2428253	-0.20	-0.16	0.11	35

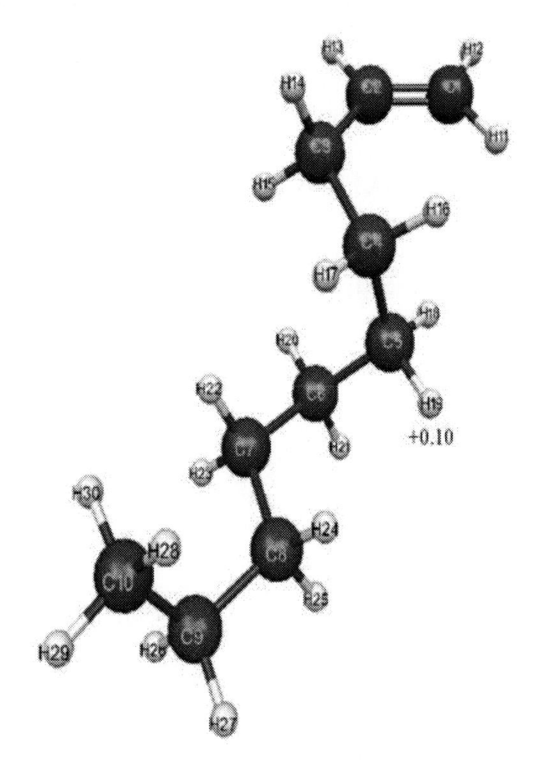

Figure 2.1.10. Geometric and electronic structure of the molecule of dekene-1 (E_0= -1023184 kDg/mol, E_{el}= -2428253 kDg/mol).

The maximum charge on atom of the hydrogen -qH+ of all under study linear olefins of cationic polymerization is +0,10 (±0.01) (within of a mistake of the method AB INITIO). pKa=36(±1) was determined by formula pKa=49,4-134,61*q_{max}^{H+}. Linear olefins possess alike and low acid power. Acid power does not depend from natures and degree of branching of olefin. Efficiency of the process of cationic polymerization and quality of the polymer does not depend from acid power of monomer.

Quantum-chemical calculation of linear monomer capable to cationic polymerization (ethylene, propylene, butene-1, butene-2, pentene-1, geksene-1, geptene-1, oktene-1, nonene-1, dekene-1) was made by classical method AB INITIO in base 6-311G** with optimization of the geometries by all parameters by standard gradient method. Optimized geometric and electronic structure of these compounds was received. Acid power was theoretically evaluated. All under investigation olefins possess alike and low acid power. Results of the calculation comparable with results got by the method CNDO/2 (pKa ≈ 43) [13]. Both methods refer linear olefins to class very weak H-Acids (pKa>14).

2.2. Calculation by Method MNDO

Quantum-chemical calculation of linear monomer (olefins) of cationic polymerization was made by method MNDO [12] with optimization of the geometries by all parameters by standard gradient method in approach the insulated molecule in gas phase. The expression pKa = 49,4-134,61·q_{max}^{H+} [6] (where pKa - an universal factor to acidity, $q_{H^+}^{\max}$ - a maximum positive charge on atom of the hydrogen) used for the estimation of acid power of linear olefins. Program MacMolPlt was used for visual presentation of the model of the molecule [14].

Results and Discussion

Geometric and electronic structure of linear olefins: ethylene, propylene, butene-1, butene-2, pentene-1, geksene-1, geptene-1, oktene-1, nonene-1, dekene-1, their quantum-chemical parameters (E_0- a general energy of the molecular system, $E_{of\ bond}$ – a total energy of all bonds of olefin, D- a dipole moment, q_H- charges on α- and β- carbon atom of olefin, q_{max}^{H+-}- a maximum positive charge on atom of the hydrogen of the monomer) and pKa – an universal factor to acidity- presented in tables 2.2.1-2.2.11 and on drawings 2.2.1-2.2.10.

Table 2.2.1. Optimized lengths of bonds, valency corners, charges of atoms of the molecule of ethylene

Length of bond	R, A	Valency corners	Grad	Atom	Charge on the atom of molecule
C(2)-C(1)	1.33	C(2)-C(1)-H(3)	123	C(1)	-0.08
H(3)-C(1)	1.09	C(2)-C(1)-H(4)	123	C(2)	-0.08
H(4)-C(1)	1.09	C(1)-C(2)-H(5)	123	H(3)	+0.04
H(5)-C(2)	1.09	C(1)-C(2)-H(6)	123	H(4)	+0.04
H(6)-C(2)	1.09			H(5)	+0.04
				H(6)	+0.04

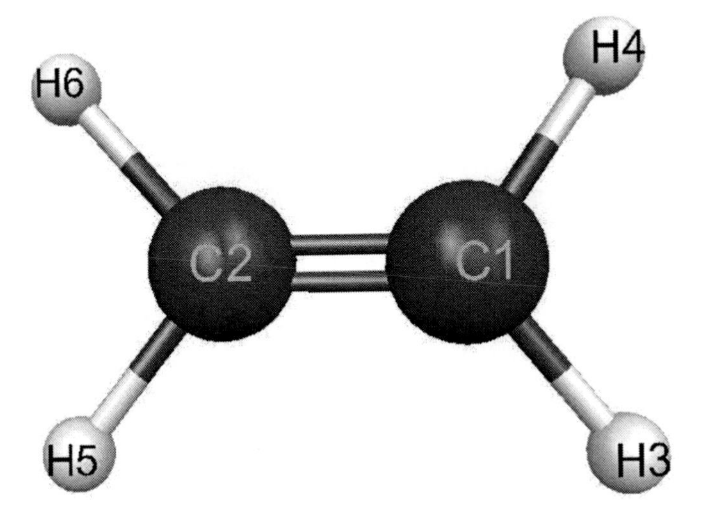

Figure 2.2.1. Geometric and electronic structure of molecule of ethylene.(E_0 = -30030 kDg/mol, E_{el} = -71497 kDg/mol).

Table 2.2.2. Optimized lengths of bonds, valency corners, charges of atoms of the molecule of propylene

Length of bond	R, A	Valency corners	Grad	Atom	Charge on the atom of molecule
C(2)-C(1)	1.34	C(1)-C(2)-C(3)	127	C(1)	-0.05
C(3)-C(2)	1.50	C(2)-C(1)-H(4)	124	C(2)	-0.13
H(4)-C(1)	1.09	C(2)-C(1)-H(5)	122	C(3)	+0.07
H(5)-C(1)	1.09	C(1)-C(2)-H(6)	119	H(4)	+0.04
H(6)-C(2)	1.10	C(2)-C(3)-H(7)	110	H(5)	+0.04
H(7)-C(3)	1.11	C(2)-C(3)-H(8)	110	**H(6)**	**+0.05**
H(8)-C(3)	1.11	C(2)-C(3)-H(9)	113	H(7)	0.00
H(9)-C(3)	1.11			H(8)	0.00
				H(9)	-0.01

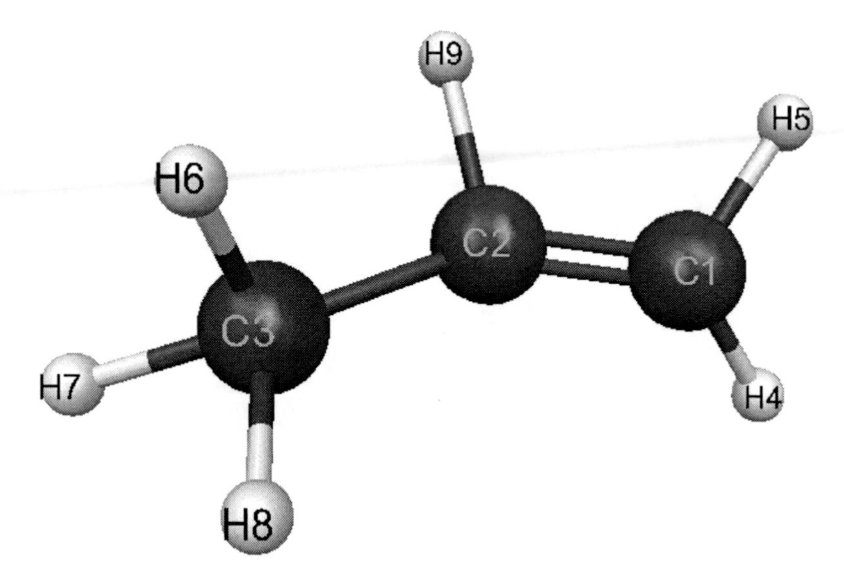

Figure 2.2.2. Geometric and electronic structure of the molecule of propylene (E_0= -45123 kDg/mol, E_{el}= -133646 kDg/mol).

Table 2.2.3.Optimized lengths of bonds, valency corners, charges of atoms of the molecule of butene–1

Length of bond	R, A	Valency corners	Grad	Atom	Charge on the atom of molecule
C(2)-C(1)	1.34	C(1)-C(2)-C(3)	127	C(1)	-0.06
C(3)-C(2)	1.50	C(2)-C(3)-C(4)	114	C(2)	-0.12
C(4)-C(3)	1.53	C(2)-C(1)-H(5)	122	C(3)	+0.02
H(5)-C(1)	1.09	C(2)-C(1)-H(6)	124	C(4)	+0.03
H(6)-C(1)	1.09	C(1)-C(2)-H(7)	119	H(5)	+0.04
H(7)-C(2)	1.10	C(2)-C(3)-H(8)	111	H(6)	+0.04
H(8)-C(3)	1.11	C(2)-C(3)-H(9)	108	**H(7)**	**+0.05**
H(9)-C(3)	1.12	C(3)-C(4)-H(10)	110	H(8)	0.00
H(10)-C(4)	1.11	C(3)-C(4)-H(11)	112	H(9)	+0.01
H(11)-C(4)	1.11	C(3)-C(4)-H(12)	112	H(10)	-0.01
H(12)-C(4)	1.11			H(11)	0.00
				H(12)	0.00

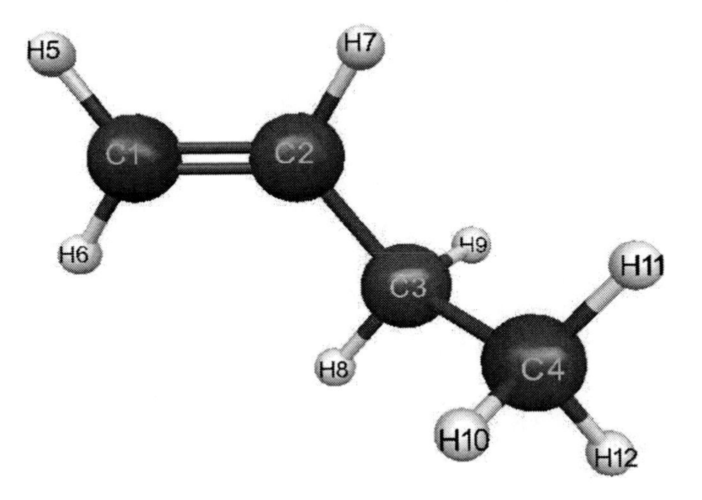

Figure 2.2.3. Geometric and electronic structure of the molecule of butene–1 (E_0= -60193 kDg/mol, E_{el}= -208529 kDg/mol).

Table 2.2.4. Optimized lengths of bonds, valency corners, charges of atoms of the molecule of butene−2

Length of bond	R, A	Valency corners	Grad	Atom	Charge on the atom of molecule
C(2)-C(1)	1.50	C(1)-C(2)-C(3)	126	C(1)	+0.06
C(3)-C(2)	1.35	C(2)-C(3)-C(4)	126	C(2)	-0.10
C(4)-C(3)	1.50	C(2)-C(1)-H(5)	113	C(3)	-0.10
H(5)-C(1)	1.11	C(2)-C(1)-H(6)	110	C(4)	+0.06
H(6)-C(1)	1.11	C(2)-C(1)-H(7)	110	H(5)	0.00
H(7)-C(1)	1.11	C(1)-C(2)-H(8)	113	H(6)	0.00
H(8)-C(2)	1.10	C(2)-C(3)-H(9)	120	H(7)	0.00
H(9)-C(3)	1.10	C(3)-C(4)-H(10)	110	H(8)	+0.04
H(10)-C(4)	1.11	C(3)-C(4)-H(11)	110	H(9)	+0.05
H(11)-C(4)	1.11	C(3)-C(4)-H(12)	113	H(10)	0.00
H(12)-C(4)	1.11			H(11)	0.00
				H(12)	0.00

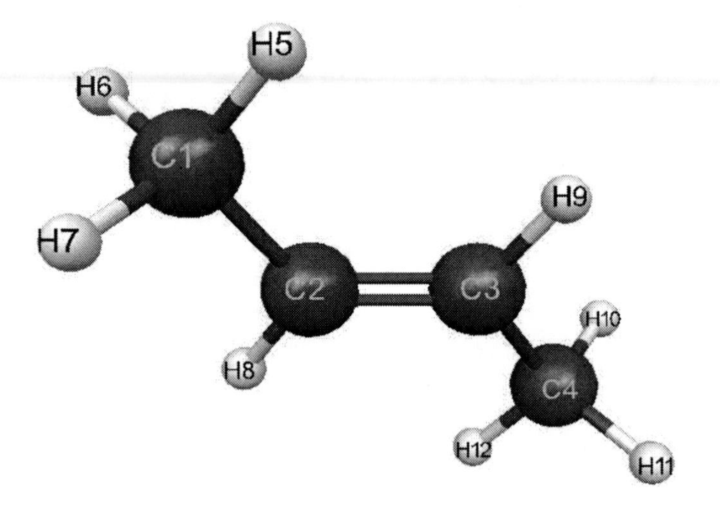

Figure 2.2.4. Geometric and electronic structure of the molecule of butene−2(E0= -60214 kDg/mol, Eel = -206456 kDg/mol).

Table 2.2.5. Optimized lengths of bonds, valency corners, charges of atoms of the molecule of pentene -1

Length of bond	R, A	Valency corners	Grad	Atom	Charge on the atom of molecule
C(2)-C(1)	1.34	C(1)-C(2)-C(3)	127	C(1)	-0.05
C(3)-C(2)	1.50	C(2)-C(3)-C(4)	113	C(2)	-0.12
C(4)-C(3)	1.54	C(3)-C(4)-C(5)	115	C(3)	+0.03
C(5)-C(4)	1.53	C(2)-C(1)-H(6)	124	C(4)	-0.02
H(6)-C(1)	1.09	C(2)-C(1)-H(7)	122	C(5)	+0.03
H(7)-C(1)	1.09	C(1)-C(2)-H(8)	119	H(6)	+0.04
H(8)-C(2)	1.10	C(2)-C(3)-H(9)	110	H(7)	+0.04
H(9)-C(3)	1.11	C(2)-C(3)-H(10)	108	**H(8)**	**+0.05**
H(10)-C(3)	1.12	C(3)-C(4)-H(11)	109	H(9)	0.00
H(11)-C(4)	1.11	C(3)-C(4)-H(12)	109	H(10)	+0.01
H(12)-C(4)	1.11	C(4)-C(5)-H(13)	112	H(11)	+0.01
H(13)-C(5)	1.11	C(4)-C(5)-H(14)	112	H(12)	+0.01
H(14)-C(5)	1.11	C(4)-C(5)-H(15)	110	H(13)	-0.01
H(15)-C(5)	1.11			H(14)	-0.01
				H(15)	-0.01

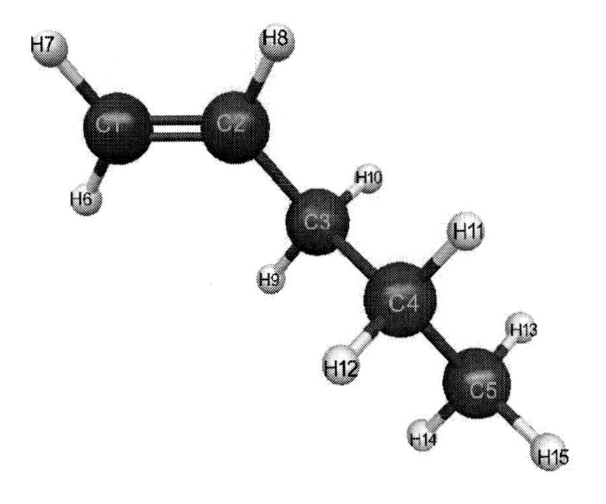

Figure 2.2.5. Geometric and electronic structure of the molecule of pentene -1 (E_0= -75261 kDg/mol, E_{el}= -292388 kDg/mol).

Table 2.2.6. Optimized lengths of bonds, valency corners, charges of atoms of the molecule of geksene-1

Length of bond	R, A	Valency corners	Grad	Atom	Charge on the atom of molecule
C(2)-C(1)	1.34	C(1)-C(2)-C(3)	127	C(1)	-0.04
C(3)-C(2)	1.50	C(2)-C(3)-C(4)	116	C(2)	-0.12
C(4)-C(3)	1.54	C(3)-C(4)-C(5)	118	C(3)	+0.03
C(5)-C(4)	1.54	C(4)-C(5)-C(6)	117	C(4)	0.00
C(6)-C(5)	1.53	C(2)-C(1)-H(7)	124	C(5)	-0.02
H(7)-C(1)	1.09	C(2)-C(1)-H(8)	122	C(6)	+0.03
H(8)-C(1)	1.09	C(1)-C(2)-H(9)	119	H(7)	+0.04
H(9)-C(2)	1.10	C(2)-C(3)-H(10)	110	H(8)	+0.04
H(10)-C(3)	1.12	C(2)-C(3)-H(11)	109	**H(9)**	**+0.05**
H(11)-C(3)	1.11	C(3)-C(4)-H(12)	108	H(10)	0.00
H(12)-C(4)	1.12	C(3)-C(4)-H(13)	108	H(11)	+0.01
H(13)-C(4)	1.12	C(4)-C(5)-H(14)	110	H(12)	0.00
H(14)-C(5)	1.11	C(4)-C(5)-H(15)	107	H(13)	0.00
H(15)-C(5)	1.12	C(5)-C(6)-H(16)	112	H(14)	+0.01
H(16)-C(6)	1.11	C(5)-C(6)-H(17)	112	H(15)	0.00
H(17)-C(6)	1.11	C(5)-C(6)-H(18)	110	H(16)	-0.01
H(18)-C(6)	1.11			H(17)	-0.01
				H(18)	-0.01

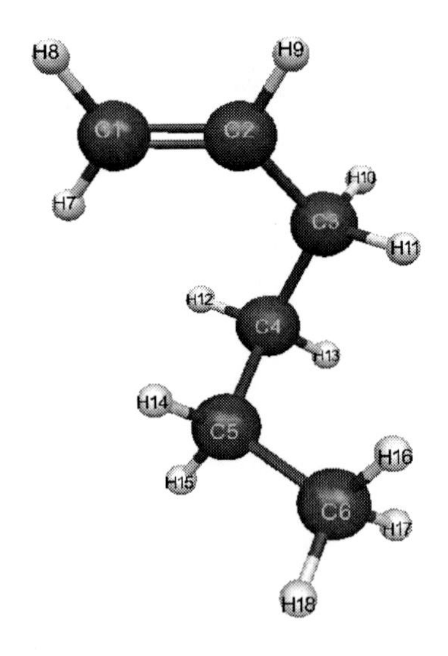

Figure 2.2.6. Geometric and electronic structure of the molecule of geksene-1 (E_0= -90325 kDg/mol, E_{el}= -390652 kDg/mol).

Table 2.2.7. Optimized lengths of bonds, valency corners, charges of atoms of the molecule of geptene-1

Length of bond	R, A	Valency corners	Grad	Atom	Charge on the atom of molecule
C(2)-C(1)	1.34	C(1)-C(2)-C(3)	127	C(1)	-0.04
C(3)-C(2)	1.50	C(2)-C(3)-C(4)	116	C(2)	-0.12
C(4)-C(3)	1.54	C(3)-C(4)-C(5)	118	C(3)	+0.03
C(5)-C(4)	1.54	C(4)-C(5)-C(6)	118	C(4)	0.00
C(6)-C(5)	1.54	C(5)-C(6)-C(7)	117	C(5)	0.00
C(7)-C(6)	1.53	C(2)-C(1)-H(8)	124	C(6)	-0.02
H(8)-C(1)	1.09	C(2)-C(1)-H(9)	122	C(7)	+0.03
H(9)-C(1)	1.09	C(1)-C(2)-H(10)	119	H(8)	+0.04
H(10)-C(2)	1.10	C(2)-C(3)-H(11)	110	H(9)	+0.04
H(11)-C(3)	1.12	C(2)-C(3)-H(12)	109	H(10)	+0.05
H(12)-C(3)	1.11	C(3)-C(4)-H(13)	108	H(11)	0.00
H(13)-C(4)	1.12	C(3)-C(4)-H(14)	108	H(12)	+0.01
H(14)-C(4)	1.11	C(4)-C(5)-H(15)	109	H(13)	0.00
H(15)-C(5)	1.11	C(4)-C(5)-H(16)	107	H(14)	0.00
H(16)-C(5)	1.12	C(5)-C(6)-H(17)	110	H(15)	0.00
H(17)-C(6)	1.11	C(5)-C(6)-H(18)	107	H(16)	0.00
H(18)-C(6)	1.12	C(6)-C(7)-H(19)	112	H(17)	+0.01
H(19)-C(7)	1.11	C(6)-C(7)-H(20)	112	H(18)	0.00
H(20)-C(7)	1.11	C(6)-C(7)-H(21)	110	H(19)	-0.01
H(21)-C(7)	1.11			H(20)	-0.01
				H(21)	-0.01

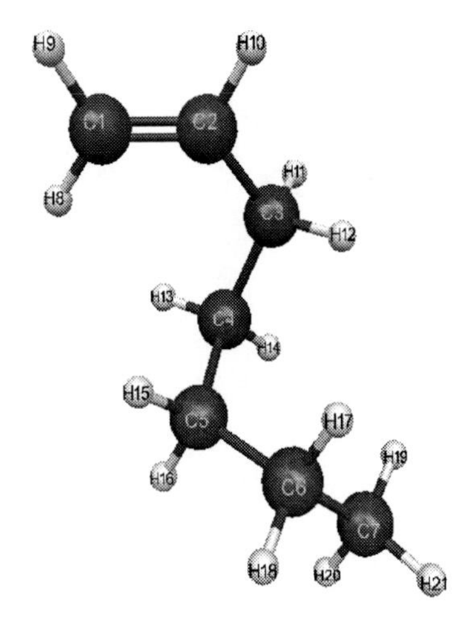

Figure 2.2.7. Geometric and electronic structure of the molecule of geptene-1 (E_0= -105390 kDg/mol, E_{el}= -493655 kDg/mol).

Table 2.2.8. Optimized lengths of bonds, valency corners, charges of atoms of the molecule of oktene-1

Length of bond	R, A	Valency corners	Grad	Atom	Charge on the atom of molecule
C(2)-C(1)	1.34	C(1)-C(2)-C(3)	127	C(1)	-0.04
C(3)-C(2)	1.51	C(2)-C(3)-C(4)	113	C(2)	-0.11
C(4)-C(3)	1.54	C(3)-C(4)-C(5)	114	C(3)	+0.03
C(5)-C(4)	1.54	C(4)-C(5)-C(6)	114	C(4)	0.00
C(6)-C(5)	1.54	C(5)-C(6)-C(7)	114	C(5)	-0.01
C(7)-C(6)	1.54	C(6)-C(7)-C(8)	115	C(6)	-0.01
C(8)-C(7)	1.53	C(2)-C(1)-H(9)	124	C(7)	-0.02
H(9)-C(1)	1.09	C(2)-C(1)-H(10)	122	C(8)	+0.03
H(10)-C(1)	1.09	C(1)-C(2)-H(11)	119	H(9)	+0.04
H(11)-C(2)	1.10	C(2)-C(3)-H(12)	110	H(10)	+0.04
H(12)-C(3)	1.11	C(2)-C(3)-H(13)	108	H(11)	+0.05
H(13)-C(3)	1.12	C(3)-C(4)-H(14)	109	H(12)	0.00
H(14)-C(4)	1.11	C(3)-C(4)-H(15)	109	H(13)	+0.01
H(15)-C(4)	1.11	C(4)-C(5)-H(16)	109	H(14)	+0.01
H(16)-C(5)	1.11	C(4)-C(5)-H(17)	109	H(15)	+0.01
H(17)-C(5)	1.11	C(5)-C(6)-H(18)	109	H(16)	0.00
H(18)-C(6)	1.11	C(5)-C(6)-H(19)	109	H(17)	0.00
H(19)-C(6)	1.11	C(6)-C(7)-H(20)	109	H(18)	0.00
H(20)-C(7)	1.11	C(6)-C(7)-H(21)	109	H(19)	0.00
H(21)-C(7)	1.11	C(7)-C(8)-H(22)	112	H(20)	0.00
H(22)-C(8)	1.11	C(7)-C(8)-H(23)	110	H(21)	0.00
H(23)-C(8)	1.11	C(7)-C(8)-H(24)	112	H(22)	-0.01
H(24)-C(8)	1.11			H(23)	-0.01
				H(24)	-0.01

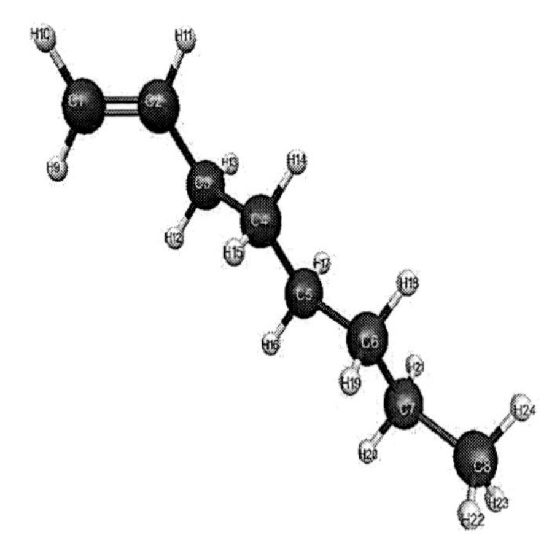

Figure 2.2.8. Geometric and electronic structure of the molecule of oktene-1 (E_0= -120465 kDg/mol, E_{el}= -584863 kDg/mol).

Table 2.2.9. Optimized lengths of bonds, valency corners, charges of atoms of the molecule of nonene-1

Length of bond	R, A	Valency corners	Grad	Atom	Charge on the atom of molecule
C(1)-C(2)	1,32			C1	-0.20
C(2)-C(3)	1,51	C(3)C(2)C(1)	128	C2	-0.16
C(3)-C(4)	1,53	C(4)C(3)C(2)	117	C3	-0.13
C(4)-C(5)	1,53	C(5)C(4)C(3)	116	C4	-0.22
C(5)-C(6)	1,53	C(6)C(5)C(4)	116	C5	-0.18
C(6)-C(7)	1,53	C(7)C(6)C(5)	116	C6	-0.20
C(7)-C(8)	1,53	C(8)C(7)C(6)	116	C7	-0.19
C(8)-C(9)	1,52	C(9)C(8)C(7)	114	C8	-0.19
C(1)-H(10)	1,08	H(10)C(1)C(2)	122	C9	-0.24
C(1)-H(11)	1,08	H(11)C(1)C(2)	120	H10	0.10
C(2)-H(12)	1,08	H(12)C(2)C(3)	114	H11	0.10
C(3)-H(13)	1,08	H(13)C(3)C(2)	107	H12	0.09
C(3)-H(14)	1,08	H(14)C(3)C(2)	108	H13	0.11
C(4)-H(15)	1,08	H(15)C(4)C(3)	109	H14	0.10
C(4)-H(16)	1,08	H(16)C(4)C(3)	107	H15	0.10
C(5)-H(17)	1,08	H(17)C(5)C(4)	109	H16	0.10
C(5)-H(18)	1,08	H(18)C(5)C(4)	107	H17	0.10
C(6)-H(19)	1,08	H(19)C(6)C(5)	109	H18	0.09
C(6)-H(20)	1,08	H(20)C(6)C(5)	107	H19	0.10
C(7)-H(21)	1,08	H(21)C(7)C(6)	109	H20	0.10
C(7)-H(22)	1,08	H(22)C(7)C(6)	107	H21	0.09
C(8)-H(23)	1,08	H(23)C(8)C(7)	110	H22	0.09
C(8)-H(24)	1,08	H(24)C(8)C(7)	107	H23	0.09
C(9)-H(25)	1,08	H(25)C(9)C(8)	111	H24	0.09
C(9)-H(26)	1,08	H(26)C(9)C(8)	110	H25	0.08
C(9)-H(27)	1,08	H(27)C(9)C(8)	111	H26	0.09
				H27	0.09

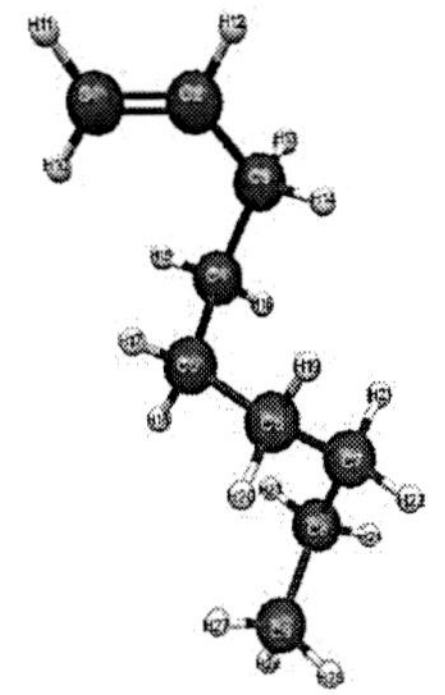

Figure 2.2.9. Geometric and electronic structure of the molecule of nonene-1 (E0= -135522 kDg/mol, Eel = -716957 kDg/mol).

Table 2.2.10. Optimized lengths of bonds, valency corners, charges of atoms of the molecule of dekene-1

Length of bond	R, A	Valency corners	Grad	Atom	Charge on the atom of molecule
C(2)-C(1)	1.34	C(1)-C(2)-C(3)	127	C(1)	-0.04
C(3)-C(2)	1.51	C(2)-C(3)-C(4)	113	C(2)	-0.11
C(4)-C(3)	1.54	C(3)-C(4)-C(5)	114	C(3)	+0.03
C(5)-C(4)	1.54	C(4)-C(5)-C(6)	114	C(4)	0.00
C(6)-C(5)	1.54	C(5)-C(6)-C(7)	114	C(5)	0.00
C(7)-C(6)	1.54	C(6)-C(7)-C(8)	114	C(6)	0.00
C(8)-C(7)	1.54	C(7)-C(8)-C(9)	114	C(7)	-0.01
C(9)-C(8)	1.54	C(8)-C(9)-C(10)	115	C(8)	-0.01
C(10)-C(9)	1.53	C(2)-C(1)-H(11)	122	C(9)	-0.02
H(11)-C(1)	1.09	C(2)-C(1)-H(12)	124	C(10)	+0.03
H(12)-C(1)	1.09	C(1)-C(2)-H(13)	119	H(11)	+0.04
H(13)-C(2)	1.10	C(2)-C(3)-H(14)	108	H(12)	+0.04
H(14)-C(3)	1.12	C(2)-C(3)-H(15)	110	**H(13)**	**+0.05**
H(15)-C(3)	1.11	C(3)-C(4)-H(16)	109	H(14)	+0.01
H(16)-C(4)	1.11	C(3)-C(4)-H(17)	109	H(15)	0.00
H(17)-C(4)	1.11	C(4)-C(5)-H(18)	109	H(16)	+0.01
H(18)-C(5)	1.11	C(4)-C(5)-H(19)	109	H(17)	+0.01
H(19)-C(5)	1.11	C(5)-C(6)-H(20)	109	H(18)	0.00
H(20)-C(6)	1.11	C(5)-C(6)-H(21)	109	H(19)	0.00
H(21)-C(6)	1.11	C(7)-C(8)-H(22)	109	H(20)	0.00
H(22)-C(8)	1.11	C(7)-C(8)-H(23)	109	H(21)	0.00
H(23)-C(8)	1.11	C(6)-C(7)-H(24)	109	H(22)	0.00
H(24)-C(7)	1.11	C(6)-C(7)-H(25)	109	H(23)	0.00
H(25)-C(7)	1.11	C(8)-C(9)-H(26)	109	H(24)	0.00
H(26)-C(9)	1.11	C(8)-C(9)-H(27)	110	H(25)	0.00
H(27)-C(9)	1.11	C(9)-C(10)-H(28)	112	H(26)	0.00
H(28)-C(10)	1.11	C(9)-C(10)-H(29)	110	H(27)	0.00
H(29)-C(10)	1.11	C(9)-C(10)-H(30)	112	H(28)	-0.01
H(30)-C(10)	1.11			H(29)	-0.01
				H(30)	-0.01

Table 2.2.11. General energy -E_0, energy of electron – E_{el}, charges on atoms - q_H, an universal factor to acidity of olefins

№п/п	Olefin	-E0 (kDg/mol)	-Eel (kDg/mol)	qCα	qCβ	qmax H+	pKa
1.	ethylene, propylene,	-30030	-71497	-	-0.08	0.04	36
2.	butene–1, butene -2,	-45123	-	0.08	-0.13	0.05	35
3.	pentene -1, geksene-1,	-60193	133646	-	-0.12	0.05	35
4.	geptene-1, oktene-1,	-60214	-	0.05	-0.10	0.05	35
5.	nonene-1, dekene-1	-75261	208529	-	-0.12	0.05	35
6.		-90325	-	0.06	-0.12	0.05	35
7.		-105390	206456	-	-0.12	0.05	35
8.		-120465	-	0.10	-0.11	0.05	35
9.		-135522	292388	-	-0.11	0.05	35
10.		-150600	-	0.05	-0.11	0.05	35
			390652	-			
			-	0.04			
			493655	-			
			-	0.04			
			584863	-			
			-	0.04			
			716957	-			
			-	0.04			
			805076	-			
				0.04			

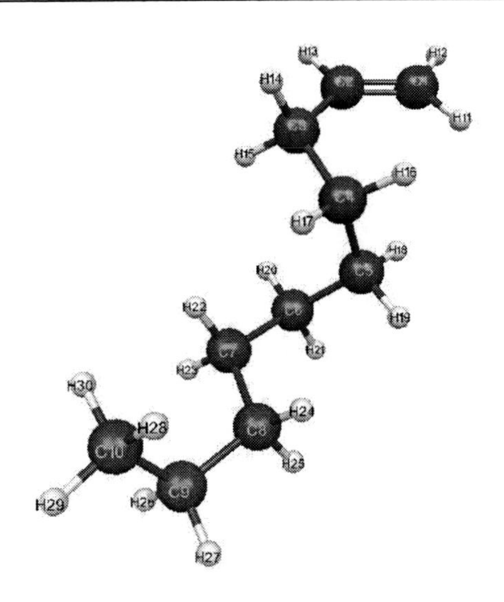

Figure 2.2.10. Geometric and electronic structure of the molecule of dekene-1. (E_0= -150600 kDg/mol, E_{el}= -805076 kDg/mol).

The maximum charge on atom of the hydrogen -qH+ of all under study linear olefins of cationic polymerization is +0,05(±0.01). pKa=35(±1) was determined by formula pKa=42,11-147,18*q_{max}^{H+}[9]. Linear olefins possess alike and low acid power. Acid power does not depend from natures and degree of branching of olefin. Efficiency of the process of cationic polymerization and quality of the polymer does not depend from acid power of monomer.

Quantum-chemical calculation of linear monomer capable to cationic polymerization (ethylene, propylene, butene-1, butene-2, pentene-1, geksene-1, geptene-1, oktene-1, nonene-1, dekene-1) was made by classical method MNDO with optimization of the geometries by all parameters by standard gradient method. Optimized geometric and electronic structure of these compounds was received. Acid power was theoretically evaluated. All under investigation olefins possess alike and low acid power pKa=+36 (±). Results of the calculation comparable with results got by the method CNDO/2 [13].

Chapter 3

RESULTS OF QUANTUM-CHEMICAL CALCULATIONS OF BRANCHED OUT OLEFINS BY METHODS AB INITIO AND MNDO

3.1. QUANTUM-CHEMICAL CALCULATION OF LINEAR OLEFINS OF CATIONIC POLYMERIZATION, BRANCHED OUT IN A-POSITION IN RELATION TO DOUBLE BOND

3.1.1. Calculation by Method AB INITIO

The Aim of this chapter is quantum-chemical calculation of molecules of monomers of cationic polymerization, branched out in α-position in relation to double bond (3-methylbutene-1, 3-methylpentene-1, 3-ethylpentene-1, 3,3-di methylbutene-1) by method AB INITIO in base 6-311G ** with optimization of the geometries by all parameters by standard gradient method in approach the insulated molecule in gas phase and a theoretical estimation of their acid force. The Method is built in in PC GAMESS [12.] Program MacMolPlt was used for visual presentation of the model of the molecule [14].

RESULTS AND DISCUSSION

Optimized geometric and electronic structures, the general energy and electronic energy of molecules (3-methylbutene-1, 3-methylpentene-1, 3-ethylpentene-1, 3,3-dimethylbutene-1) are received by method AB INITIO in basis 6-311G ** and are shown in tables 3.1.1-3.1.8 and on drawings 3.1.1-3.1.7. pKa is certain under the formula pKa=49,4-134,61*q_{max}^{H+} [9]. R=0.96, R - the factor of correlation, q_{max}^{H+}=0,1.

Thus, quantum-chemical calculation of molecules (3-methylbutene-1, 3-methylpentene-1, 3-ethylpentene-1, 3,3-dimethylbutene-1) for the first time it is executed by method AB INITIO in base 6-311G **. Optimized geometric and electronic structures of these compounds were received. Acid power was theoretically evaluated. We have shown that all of them possess identical acid force. pKa=35. We have positioned, that monomers of cationic polymerization, branched out in α-position in relation to double bond (3-methylbutene-1, 3-methylpentene-1, 3-ethylpentene-1, 3,3-dimethylbutene-1) concern to a class of very weak H-acids (pKa> 14).

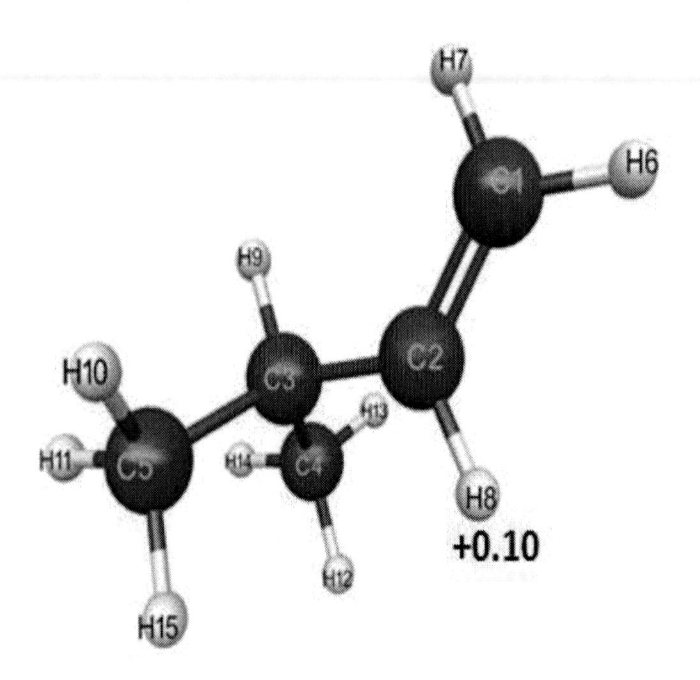

Figure 3.1.1. Geometric and electronic structure of the molecule of 3-methylbutene-1 (E0= -511068 kDg/mol.).

Table 3.1.1.Optimized lengths of bonds, valency corners, charges of atoms of the molecule of 3-methylbutene-1

Length of bond	R, A	Valency corners	Grad	Atom	Charge (by Milliken)
C(1)-C(2)	1.34	C(1)-C(2)-C(3)	128	C1	-0.20
C(2)-C(3)	1.51	C(2)-C(3)-C(4)	114	C2	-0.13
C(3)-C(4)	1.53	C(2)-C(3)-C(5)	110	C3	-0.16
C(3)-C(5)	1.54	C(2)-C(1)-H(6)	121	C4	-0.24
C(1)-H(6)	1.08	C(2)-C(1)-H(7)	121	C5	-0.23
C(1)-H(7)	1.08	C(3)-C(2)-H(8)	116	H6	0.10
C(2)-H(8)	1.08	C(2)-C(3)-H(9)	118	H7	0.10
C(3)-H(9)	1.09	C(3)-C(4)-H(12)	106	**H8**	**0.10**
C(4)-H(12)	1.09	C(3)-C(4)-H(13)	111	H9	0.10
C(4)-H(13)	1.09	C(3)-C(4)-H(14)	108	H10	0.09
C(4)-H(14)	1.09	C(3)-C(5)-H(10)	111	H11	0.09
C(5)-H(10)	1.09	C(3)-C(5)-H(11)	108	H12	0.09
C(5)-H(11)	1.09	C(3)-C(5)-C(15)	108	H13	0.10
C(5)-H(15)	1.09			H14	0.09
				H15	0.09

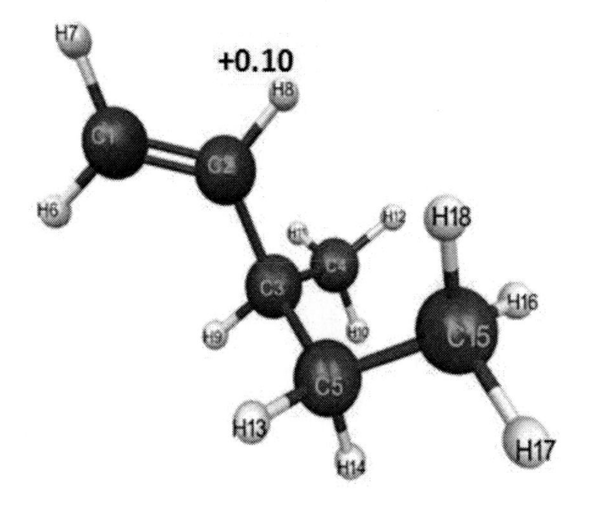

Figure 3.1.2. Geometric and electronic structure of the molecule of 3-methylpentene-1 (E0= - 612421 kDg/mol.)

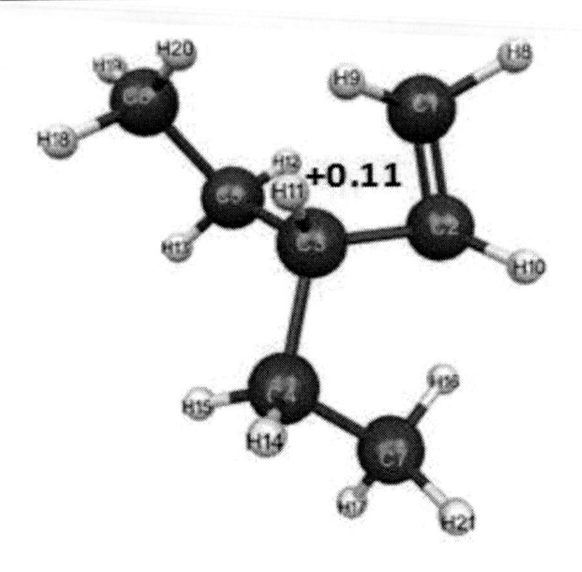

Figure 3.1.3. Geometric and electronic structure of the molecule of 3-ethylpentene-1. (E0= -715495 kDg/mol).

Table 3.1.2. Optimized lengths of bonds, valency corners, charges of atoms of the molecule of 3-methylpentene-1

Length of bond	R, A	Valency corners	Grad	Atom	Charge (by Milliken)
C(1)-C(2)	1.32	C(1)-C(2)-C(3)	128	C1	-0.20
C(2)-C(3)	1.51	C(2)-C(3)-C(4)	114	C2	-0.13
C(3)-C(4)	1.53	C(3)-C(5)-C(6)	114	C3	-0.15
C(3)-C(5)	1.54	C(3)-C(4)-C(6)	109	C4	-0.21
C(4)-C(6)	1.53	C(2)-C(1)-H(8)	114	H5	-0.22
C(1)-H(8)	1.08	C(2)-C(1)-H(9)	123	H6	-0.23
C(1)-H(7)	1.08	C(2)-C(1)-H(9)	116	C7	0.10
C(2)-H(9)	1.08	C(2)-C(3)-H(10)	118	**H8**	**0.10**
C(3)-H(10)	1.09	C(3)-C(4)-H(14)	106	H9	0.09
C(4)-H(14)	1.09	C(3)-C(4)-H(15)	111	H10	0.10
C(4)-H(15)	1.08	C(3)-C(5)-H(11)	108	H11	0.09
C(5)-H(11)	1.09	C(3)-C(5)-H(12)	108	H12	0.10
C(5)-H(12)	1.09	C(3)-C(5)-H(13)	108	H13	0.09
C(5)-H(13)	1.09	C(4)-C(6)-H(16)	107	H14	0.10
C(6)-H(16)	1.09	C(4)-C(6)-H(17)	110	H15	0.09
C(6)-H(17)	1.09	C(4)-C(6)-H(18)	107	H16	0.09
C(6)-H(18)	1.09			H17	0.09
				H18	0.08

Table 3.1.3. Optimized lengths of bonds, valency corners, charges of atoms of the molecule of 3-ethylpentene-1

Length of bond	R, A	Valency corners	Grad	Atom	Charge (by Milliken)
C(1)-C(2)	1.32	C(1)-C(2)-C(3)	129	C1	-0.19
C(2)-C(3)	1.52	C(2)-C(3)-C(4)	110	C2	-0.12
C(3)-C(4)	1.55	C(3)-C(5)-C(6)	115	C3	-0.16
C(3)-C(5)	1.54	C(3)-C(4)-C(7)	114	C4	-0.18
C(4)-C(7)	1.53	C(5)-C(6)-H(18)	116	H5	-0.20
C(5)-C(6)	1.53	C(2)-C(1)-H(8)	121	H6	-0.22
C(1)-H(8)	1.08	C(2)-C(1)-H(9)	116	C7	-0.23
C(1)-C(9)	1.08	C(1)-C(2)-H(10)	116	H8	0.10
C(2)-H(10)	1.08	C(1)-C(2)-H(11)	118	H9	0.10
C(3)-H(11)	1.09	C(3)-C(4)-H(14)	106	H10	0.09
C(4)-H(14)	1.09	C(3)-C(4)-H(15)	110	**H11**	**+0.11**
C(4)-H(15)	1.09	C(3)-C(5)-H(12)	106	H12	0.09
C(5)-H(12)	1.09	C(3)-C(5)-H(13)	109	H13	0.09
C(5)-H(13)	1.09	C(5)-C(6)-H(18)	106	H14	0.09
C(6)-H(18)	1.09	C(5)-C(6)-H(19)	112	H15	0.09
C(6)-H(19)	1.09	C(5)-C(6)-H(20)	108	H16	0.09
C(6)-H(20)	1.09	C(4)-C(7)-H(16)	111	H17	0.09
C(7)-H(16)	1.09	C(4)-C(7)-H(17)	108	H18	0.08
C(7)-H(17)	1.09	C(4)-C(7)-H(21)	108	H19	0.09
C(7)-H(21)	1.09			H20	0.09
				H21	0.09

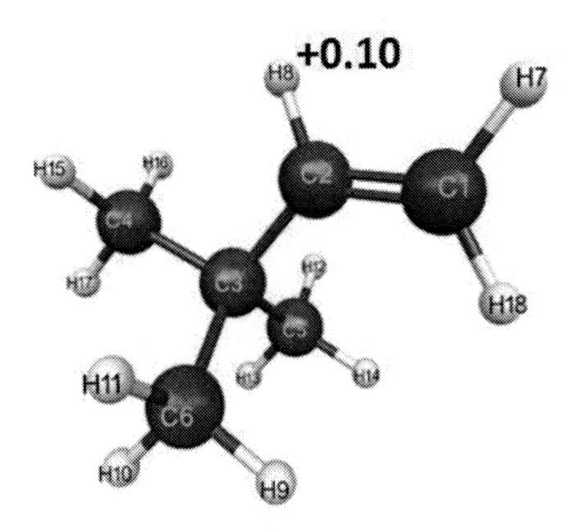

Figure 3.1.4. Geometric and electronic structure of the molecule of 3,3-dimethylbutene-1 (E0= - 714500 kDg/mol).

Table 3.1.4.Optimized lengths of bonds, valency corners, charges of atoms of the molecule of 3,3-dimethylbutene-1

Length of bond	R, A	Valency corners	Grad	Atom	Charge (by Milliken)
C(1)-C(2)	1,32	C(1)-C(2)-C(3)	128	C1	-0.19
C(2)-C(3)	1.52	C(2)-C(3)-C(4)	128	C2	-0.09
C(3)-C(4)	1.54	C(2)-C(3)-C(5)	108	C3	-0.25
C(3)-C(5)	1.54	C(3)-C(6)-H(9)	108	C4	-0.20
C(3)-C(6)	1.54	C(2)-C(1)-H(7)	113	C5	-0.20
C(1)-H(7)	1.08	C(2)-C(1)-H(18)	121	C6	-0.22
C(1)-H(18)	1.08	C(1)-C(2)-H(8)	118	H7	0.10
C(2)-H(8)	1.08	C(3)-C(4)-H(15)	112	**H8**	**0.10**
C(4)-H(15)	1.09	C(3)-C(4)-H(16)	108	H9	0.10
C(4)-H(16)	1.09	C(3)-C(4)-H(17)	108	H10	0.09
C(4)-H(17)	1.09	C(3)-C(5)-H(12)	111	H11	0.10
C(5)-H(12)	1.09	C(3)-C(5)-H(13)	108	H12	0.09
C(5)-H(13)	1.09	C(3)-C(5)-C(14)	108	H13	0.09
C(5)-H(14)	1.09	C(3)-C(6)-H(9)	111	H14	0.10
C(6)-H(9)	1.09	C(3)-C(6)-H(10)	108	H15	0.10
C(6)-H(10)	1.09	C(3)-C(6)-H(11)	108	H16	0.09
C(6)-H(11)	1.08			H17	0.09
				H18	0.10

Table 3.1.5. The General energy -E_0, the electronic energy of bond (E_{el}), the maximal charge on atom of hydrogen (q_H^{max}), the universal parameter of acidity (pKa) of monomers of cationic polymerization, branched out in α-position in relation to double bond

№	Monomers	E_0, kDg/mol	q_H^{max}	pKa
1	3-methylbutene-1	-511068	+0,10	36
2	3-methylpentene-1	-612421	+0,10	36
3	3-ethylpentene-1	-715495	+0,11	34
4	3,3-dimethylbutene-1	-714500	+0,10	36

3.1.2. Calculation by Method MNDO

The Aim of this chapter is quantum-chemical calculation of molecules of monomers of cationic polymerization, branched out in α-position in relation to double bond (3-methylbutene-1, 3-methylpentene-1, 3-ethylpentene-1, 3,3-di methylbutene-1) by method MNDO with optimization of the geometries by all parameters by standard gradient method in approach the insulated molecule in gas phase and a theoretical estimation of their acid force. The Method is built in in PC GAMESS [12.] Program MacMolPlt was used for visual presentation of the model of the molecule [14].

Results and Discussion

Optimized geometric and electronic structures, the general energy and electronic energy of molecules (3-methylbutene-1, 3-methylpentene-1, 3-ethylpentene-1, 3,3-dimethylbutene-1) are received by method MNDO and are shown in tables 3.1.6-3.1.10 and on drawings 3.1.5-3.1.8. pKa is certain under the formula pKa=42.11-147.18q_{max}^{H+} [9] (q_{max}^{H+}=+0.05 - the maximal charge on atom of hydrogen, pKa - a universal parameter of acidity Table 3.1.10) pKa=35.

Thus, quantum-chemical calculation of molecules (3-methylbutene-1, 3-methylpentene-1, 3-ethylpentene-1, 3,3-dimethylbutene-1) for the first time it is executed by method MNDO. Optimized geometric and electronic structures of these compounds were received. Acid power was theoretically evaluated. We have shown that all of them possess identical acid force. pKa=35. We have positioned, that monomers of cationic polymerization, branched out in α-position in relation to double bond (3-methylbutene-1, 3-methylpentene-1, 3-ethylpentene-1, 3,3-dimethylbutene-1) concern to a class of very weak H-acids (pKa> 14).

Table 3.1.6. Optimized lengths of bonds, valency corners, charges of atoms of the molecule of 3-methylbutene-1

Length of bond	R, A	Valency corners	Grad	Atom	Charge (by Milliken)
C(1)-C(2)	1.34	C(1)-C(2)-C(3)	126	C1	-0.06
C(2)-C(3)	1.51	C(2)-C(3)-C(4)	111	C2	-0.11
C(3)-C(4)	1.54	C(2)-C(3)-C(5)	111	C3	-0.03
C(3)-C(5)	1.54	C(2)-C(1)-H(6)	122	C4	0.04
C(1)-H(6)	1.10	C(2)-C(1)-H(7)	113	C5	0.04
C(1)-H(7)	1.10	C(1)-C(2)-H(8)	119	H6	0.04
C(2)-H(8)	1.10	C(5)-C(3)-H(9)	106	H7	0.04
C(3)-H(9)	1.12	C(3)-C(4)-H(12)	111	**H8**	**0.05**
C(4)-H(12)	1.11	C(3)-C(4)-H(13)	107	H9	0.01
C(4)-H(13)	1.12	C(3)-C(4)-H(14)	112	H10	0.00
C(4)-H(14)	1.11	C(3)-C(5)-H(10)	108	H11	-0.01
C(5)-H(10)	1.11	C(3)-C(5)-H(11)	107	H12	0.00
C(5)-H(11)	1.11	C(3)-C(5)-H(15)	108	H13	0.00
C(5)-H(15)	1.11			H14	-0.01
				H15	0.00

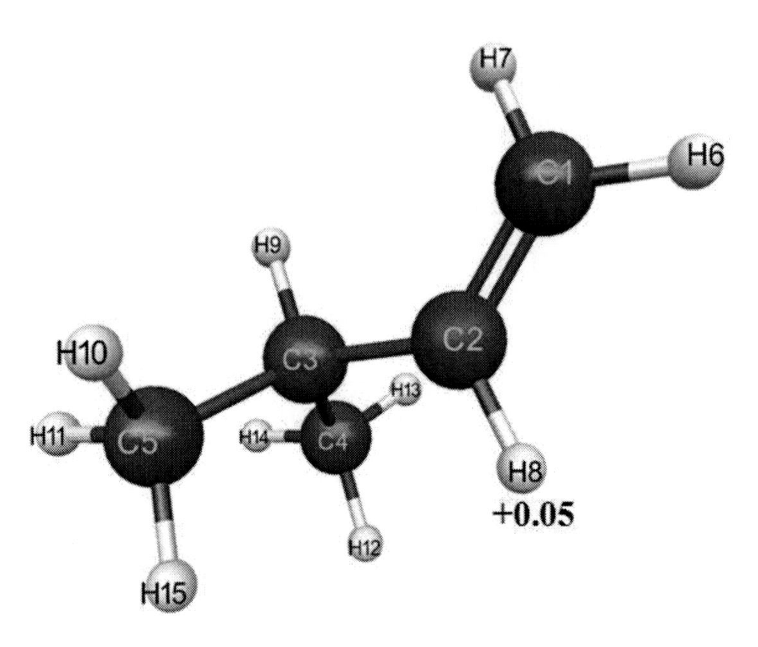

Figure 3.1.5. Geometric and electronic structure of the molecule of 3-methylbutene-1 (E0= -299348 kDg/mol.).

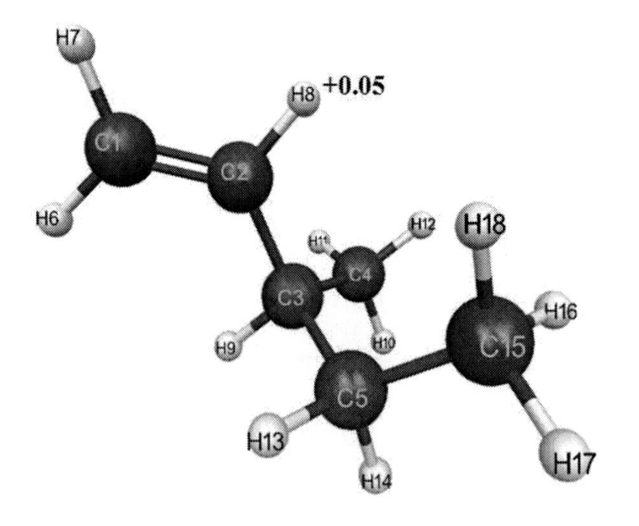

Figure 3.1.6. Geometric and electronic structure of the molecule of 3-methylpentene-1 (E0= - 399131 кДж/моль, Eel= - 91905 kDg/mol.).

Table 3.1.7. Optimized lengths of bonds, valency corners, charges of atoms of the molecule of 3-methylpentene-1

Length of bond	R, A	Valency corners	Grad	Atom	Charge (by Milliken)
C(1)-C(2)	1.34	C(1)-C(2)-C(3)	126	C1	-0.06
C(2)-C(3)	1.51	C(2)-C(3)-C(4)	111	C2	-0.11
C(3)-C(4)	1.54	C(2)-C(3)-C(5)	112	C3	-0.01
C(3)-C(5)	1.55	C(2)-C(1)-H(6)	124	C4	0.04
C(1)-H(6)	1.09	C(2)-C(1)-H(7)	113	C5	-0.01
C(1)-H(7)	1.09	C(1)-C(2)-H(8)	118	H6	0.04
C(2)-H(8)	1.10	C(5)-C(3)-H(9)	104	H7	0.04
C(3)-H(9)	1.12	C(3)-C(4)-H(10)	111	**H8**	**0.05**
C(4)-H(10)	1.11	C(3)-C(4)-H(11)	107	H9	0.01
C(4)-H(11)	1.11	C(3)-C(4)-H(12)	107	H10	-0.01
C(4)-H(12)	1.11	C(3)-C(5)-H(13)	107	H11	-0.01
C(5)-H(13)	1.12	C(3)-C(5)-H(14)	105	H12	0.00
C(5)-H(14)	1.12	C(3)-C(5)-C(15)	118	H13	0.00
C(5)-C(15)	1.53	C(5)-C(15)-H(16)	112	H14	0.00
C(15)-H(16)	1.11	C(5)-C(15)-H(17)	107	H15	0.03
C(15)-H(17)	1.11	C(5)-C(15)-H(18)	107	H16	0.00
C(15)-H(18)	1.11			H17	-0.01
				H18	0.00

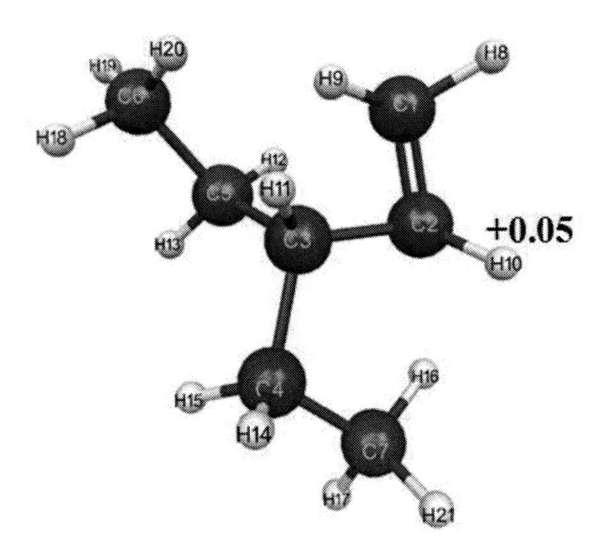

Figure 3.1.7. Geometric and electronic structure of the molecule of 3-ethylpentene-1(E0= - 104834 кДж/моль, Eel= -505826 kDg/mol).

Table 3.1.8. Optimized lengths of bonds, valency corners, charges of atoms of the molecule of 3-ethylpentene-1

Length of bond	R, A	Valency corners	Grad	Atom	Charge (by Milliken)
C(1)-C(2)	1.33	C(1)-C(2)-C(3)	86	C1	-0.06
C(2)-C(3)	1.50	C(2)-C(3)-C(4)	108	C2	-0.11
C(3)-C(4)	1.58	C(3)-C(4)-C(7)	110	C3	0.00
C(4)-C(7)	1.56	C(3)-C(5)-C(6)	108	C4	-0.01
C(3)-C(5)	1.54	C(5)-C(6)-H(18)	110	C5	-0.01
C(5)-C(6)	1.54	C(2)-C(1)-H(8)	110	C6	0.03
C(1)-H(8)	1.14	C(2)-C(1)-H(9)	110	C7	0.03
C(1)-H(9)	1.14	C(3)-C(2)-H(10)	115	H8	0.04
C(2)-H(10)	1.14	C(2)-C(3)-H(11)	93	H9	0.04
C(3)-H(11)	1.11	C(3)-C(4)-H(14)	110	**H10**	**0.05**
C(4)-H(14)	1.14	C(3)-C(4)-H(15)	110	H11	0.01
C(4)-H(15)	1.04	C(3)-C(5)-H(12)	110	H12	0.01
C(5)-H(12)	1.14	C(3)-C(5)-C(13)	110	H13	0.00
C(5)-H(13)	1.14	C(3)-C(5)-H(13)	110	H14	0.00
C(5)-H(13)	1.14	C(5)-C(6)-H(18)	110	H15	0.00
C(6)-H(18)	1.14	C(5)-C(6)-H(19)	110	H16	0.00
C(6)-H(19)	1.14	C(5)-C(6)-H(20)	110	H17	-0.01
C(6)-H(20)	1.14	C(4)-C(7)-H(16)	91	H18	0.00
C(7)-H(16)	1.04	C(4)-C(7)-H(17)	110	H19	-0.01
C(7)-H(17)	1.14	C(4)-C(7)-H(21)	110	H20	0.00
C(7)-H(21)	1.14			H21	0.00

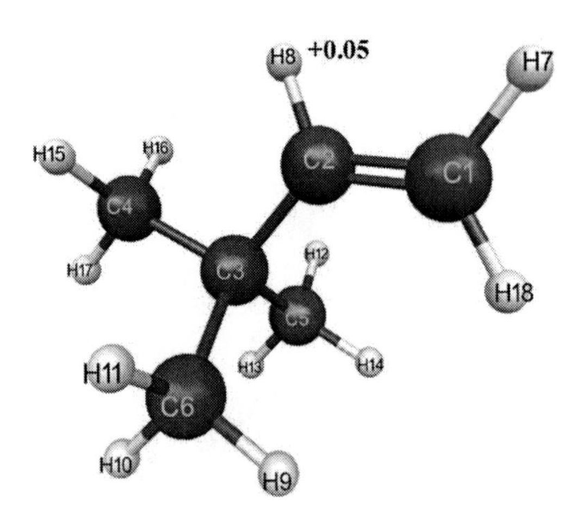

Figure 3.1.8. Geometric and electronic structure of the molecule of 3,3-dimethylbutene-1 (E0= - 408261 кДж/моль, Eel=- 90453 kDg/mol).

Table 3.1.9. Optimized lengths of bonds, valency corners, charges of atoms of the molecule of 3,3-dimethylbutene-1

Length of bond	R, A	Valency corners	Grad	Atom	Charge (by Milliken)
C(1)-C(2)	1.34	C(1)-C(2)-C(3)	128	C1	-0.06
C(2)-C(3)	1.52	C(2)-C(3)-C(4)	109	C2	-0.10
C(3)-C(4)	1.56	C(2)-C(3)-C(5)	109	C3	-0.07
C(3)-C(5)	1.56	C(3)-C(6)-H(9)	111	C4	0.05
C(3)-C(6)	1.56	C(2)-C(1)-H(7)	122	C5	0.05
C(1)-H(7)	1.10	C(2)-C(1)-H(18)	117	C6	0.05
C(1)-H(18)	1.10	C(1)-C(2)-H(8)	112	H7	0.04
C(2)-H(8)	1.11	C(3)-C(4)-H(15)	107	H8	+0.05
C(4)-H(15)	1.11	C(3)-C(4)-H(16)	107	H9	-0.01
C(4)-H(16)	1.11	C(3)-C(4)-H(17)	111	H10	-0.01
C(4)-H(17)	1.11	C(3)-C(5)-H(12)	107	H11	-0.01
C(5)-H(12)	1.11	C(3)-C(5)-H(13)	107	H12	-0.01
C(5)-H(13)	1.11	C(3)-C(5)-C(14)	112	H13	-0.01
C(5)-H(14)	1.11	C(3)-C(6)-H(9)	107	H14	-0.01
C(6)-H(9)	1.11	C(3)-C(6)-H(10)	107	H15	-0.01
C(6)-H(10)	1.11	C(3)-C(6)-H(11)	113	H16	-0.01
C(6)-(11)	1.10			H17	-0.01
				H18	0.04

The Table 3.1.10. The General energy -E0, the electronic energy of bond (Eel), the maximal charge on atom of hydrogen (q_H^{max}), the universal parameter of acidity (pKa) of monomers of cationic polymerization, branched out in α-position in relation to double bond

№	Monomer	E_0, kDg/mol	E_{el}, kDg/mol	q_H^{max}	pKa
1	3-methylbutene-1	-299348	-76150	+0,05	35
2	3-methylpentene-1	- 399131	- 91905	+0,05	35
3	3-ethylpentene-1	- 104834	- 505826	+0,05	35
4	3,3-dimethylbutene-1	- 408261	- 90453	+0,05	35

3.2. Quantum-Chemical Calculation of Olefins of Cationic Polymerization, Branched out in B -Position in Relation to Double Bond

3.2.1. Calculation by Method AB INITIO

The Aim of this chapter is quantum-chemical calculation of molecules of monomers of cationic polymerization, branched out in α-position in relation to double bond (4-methylpentene-1, 4-methylhexen-1, 4,4-dimethylpentene-1) by method AB INITIO in base 6-311G ** with optimization of the geometries by all parameters by standard gradient method in approach the insulated molecule in gas phase and a theoretical estimation of their acid force. The Method is built in in PC GAMESS [12.] Program MacMolPlt was used for visual presentation of the model of the molecule [14].

Results and Discussion

Optimized geometric and electronic structures, the general energy and electronic energy of molecules (4-methylpentene-1, 4-methylhexen-1, 4,4-dimethylpentene-1) are received by method AB INITIO in basis 6-311G ** and are shown in tables 3.2.1-3.2.3 and on drawings 3.2.1-3.2.3. pKa is certain under the formula pKa=49,4-134,61*q_{max}^{H+} [9]. q_{max}^{H+}= +0,11. pKa=35.

Thus, quantum-chemical calculation of molecules (4-methylpentene-1, 4-methylhexen-1, 4,4-dimethylpentene-1) for the first time it is executed by method AB INITIO in base 6-311G **. Optimized geometric and electronic structures of these compounds were received. Acid power was theoretically evaluated. pKa=35. We have positioned, that these molecules (4-methylpentene-1, 4-methylhexen-1, 4,4-dimethylpentene-1) concern to a class of very weak H-acids (pKa> 14).

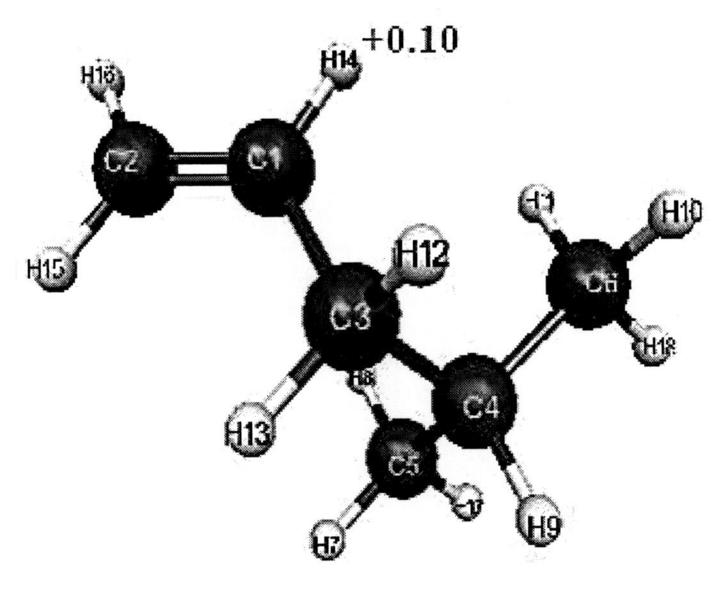

Figure 3.2.1. Geometric and electronic structure of the molecule of 4-methylpentene-1 (E0= -613876 кДж/моль, Eel= -2055339 kDg/mol).

Table 3.2.1. Optimized lengths of bonds, valency corners, charges of atoms of the molecule of 4-methylpentene-1

Length of bond	R, A	Valency corners	Grad	Atom	Charge on the atom of molecule
				C(1)	-0.17
C(1)-C(3)	1,51			C(2)	-0.18
C(2)-C(1)	1,32	C(2)-C(1)-C(3)	125	C(3)	-0.16
C(3)-C(4)	1,54	C(1)-C(3)-C(4)	116	C(4)	-0.19
C(4)-C(5)	1,53	C(3)-C(4)-C(5)	112	C(5)	-0.21
C(6)-C(4)	1,53	C(3)-C(4)-C(6)	113	C(6)	-0.23
H(7)-C(5)	1,09	C(4)-C(5)-H(7)	111	H(7)	+0.09
H(8)-C(5)	1,09	C(4)-C(5)-H(8)	111	H(8)	+0.09
H(9)-C(4)	1,09	C(3)-C(4)-H(9)	106	H(9)	+0.09
H(10)-C(6)	1,09	C(4)-C(6)-H(10)	111	H(10)	+0.09
H(11)-C(6)	1,09	C(4)-C(6)-H(11)	112	H(11)	+0.09
H(12)-C(3)	1,09	C(1)-C(3)-H(12)	118	H(12)	+0.10
H(13)-C(3)	1,09	C(1)-C(3)-H(13)	109	H(13)	+0.10
H(14)-C(1)	1,08	C(2)-C(1)-H(14)	118	H(14)	+0.10
H(15)-C(2)	1,08	C(1)-C(2)-H(15)	122	H(15)	+0.10
H(16)-C(2)	1,08	C(1)-C(2)-H(16)	122	H(16)	+0.10
H(17)-C(5)	1,09	C(4)-C(5)-H(17)	110	H(17)	+0.09
H(18)-C(6)	1,09	C(4)-C(6)-H(18)	110	H(18)	+0.09

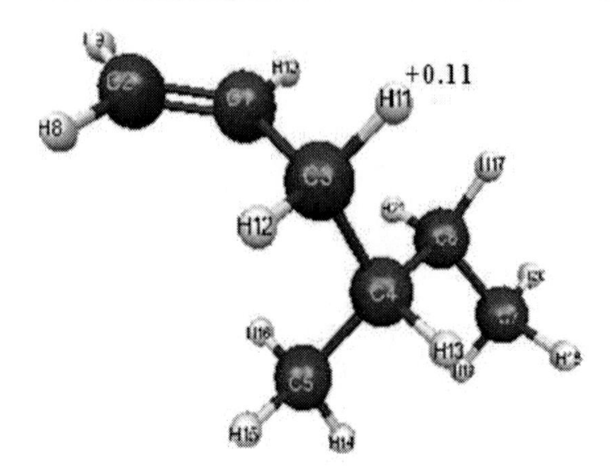

Figure 3.2.2. Geometric and electronic structure of the molecule of 4-methylhexen-1 (E0= -716197 кДж/моль, Eel= -2544480 kDg/mol).

Table 3.2.2. Optimized lengths of bonds, valency corners, charges of atoms of the molecule 4-methylhexen-1

Length of bond	R, A	Valency corners	Grad	Atom	Charge on the atom of molecule
				C(1)	-0.16
C(1)-C(3)	1.51			C(2)	-0.18
C(2)-C(1)	1.32	C(2)-C(1)-C(3)	125	C(3)	-0.15
C(3)-C(4)	1.54	C(1)-C(3)-C(4)	116	C(4)	-0.19
C(4)-C(5)	1.53	C(3)-C(4)-C(5)	111	C(5)	-0.21
C(6)-C(4)	1.54	C(3)-C(4)-C(6)	112	C(6)	-0.20
C(7)-C(6)	1.52	C(4)-C(6)-C(7)	114	C(7)	-0.23
H(8)-C(2)	1.08	C(1)-C(2)-H(8)	122	H(8)	+0.10
H(9)-C(2)	1.08	C(1)-C(2)-H(9)	122	H(9)	+0.10
H(10)-C(1)	1.08	C(2)-C(1)-H(10)	118	H(10)	+0.10
H(11)-C(3)	1.09	C(1)-C(3)-H(11)	108	**H(11)**	**+0.11**
H(12)-C(3)	1.09	C(1)-C(3)-H(12)	109	H(12)	+0.10
H(13)-C(4)	1.09	C(3)-C(4)-H(13)	105	H(13)	+0.09
H(14)-C(5)	1.09	C(4)-C(5)-H(14)	112	H(14)	+0.09
H(15)-C(5)	1.09	C(4)-C(5)-H(15)	111	H(15)	+0.09
H(16)-C(5)	1.09	C(4)-C(5)-H(16)	111	H(16)	+0.09
H(17)-C(6)	1.09	C(4)-C(6)-H(17)	109	H(17)	+0.09
H(18)-C(7)	1.09	C(6)-C(7)-H(18)	112	H(18)	+0.08
H(19)-C(7)	1.09	C(6)-C(7)-H(19)	111	H(19)	+0.09
H(20)-C(7)	1.09	C(6)-C(7)-H(20)	111	H(20)	+0.09
H(21)-C(6)	1.09	C(4)-C(6)-H(21)	109	H(21)	+0.09

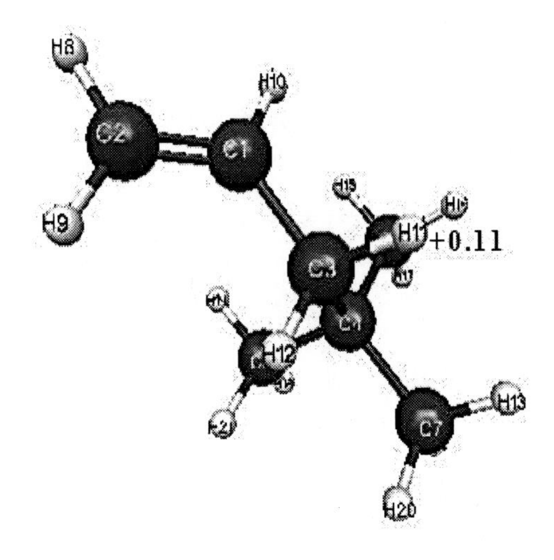

Figure 3.2.3. Geometric and electronic structure of the molecule of 4,4-dimethylpentene-1 (E0= -716202 кДж/моль, Eel= -2588309 kDg/mol).

Table 3.2.3. Optimized lengths of bonds, valency corners, charges of atoms of the molecule 4,4-dimethylpentene-1

Length of bond	R, A	Valency corners	Grad	Atom	Charge on the atom of molecule
				C(1)	-0.16
C(1)-C(3)	1,51			C(2)	-0.18
C(2)-C(1)	1,32	C(2)-C(1)-C(3)	124	C(3)	-0.12
C(3)-C(4)	1,55	C(1)-C(3)-C(4)	116	C(4)	-0.27
C(4)-C(5)	1,54	C(3)-C(4)-C(5)	111	C(5)	-0.19
C(6)-C(4)	1,54	C(3)-C(4)-C(6)	111	C(6)	-0.21
C(7)-C(4)	1,54	C(3)-C(4)-C(7)	108	C(7)	-0.20
H(8)-C(2)	1,08	C(1)-C(2)-H(8)	122	H(8)	+0.10
H(9)-C(2)	1,08	C(1)-C(2)-H(9)	122	H(9)	+0.10
H(10)-C(1)	1,08	C(2)-C(1)-H(10)	118	H(10)	+0.10
H(11)-C(3)	1,09	C(1)-C(3)-H(11)	108	**H(11)**	**+0.11**
H(12)-C(3)	1,09	C(1)-C(3)-H(12)	109	H(12)	+0.10
H(13)-C(7)	1,09	C(4)-C(7)-H(13)	111	H(13)	+0.10
H(14)-C(7)	1,09	C(4)-C(7)-H(14)	111	H(14)	+0.09
H(15)-C(6)	1,09	C(4)-C(6)-H(15)	112	H(15)	+0.09
H(16)-C(6)	1,09	C(4)-C(6)-H(16)	111	H(16)	+0.09
H(17)-C(6)	1,09	C(4)-C(6)-H(17)	110	H(17)	+0.09
H(18)-C(5)	1,09	C(4)-C(5)-H(18)	111	H(18)	+0.09
H(19)-C(5)	1,09	C(4)-C(5)-H(19)	112	H(19)	+0.10
H(20)-C(7)	1,09	C(4)-C(7)-H(20)	111	H(20)	+0.09
H(21)-C(5)	1,09	C(4)-C(5)-H(21)	111	H(21)	+0.09

Table 3.2.4. The General energy $-E_0$, the electronic energy of bond (E_{el}), the maximal charge on atom of hydrogen (q_H^{max}), the universal parameter of acidity (pKa) of monomers of cationic polymerization, branched out in β-position in relation to double bond

№	Monomer	$-E_0$ kDg/mol	$-E_{el}$ kDg/mol	q_{max}^{H+}	pKa
1	4-methylpentene-1	613876	2055339	+0,10	36
2	4-methylhexen-1	716197	2544480	+0,11	35
3	4,4-dimethylpentene-1	716202	2588309	+0,11	35

3.2.2. Calculation by Method MNDO

The Aim of this chapter is quantum-chemical calculation of molecules of (4-methylpentene-1, 4-methylhexen-1, 4,4-dimethylpentene-1) by method MNDO with optimization of the geometries by all parameters by standard gradient method in approach the insulated molecule in gas phase and a theoretical estimation of their acid force. The Method is built in in PC GAMESS [12.] Program MacMolPlt was used for visual presentation of the model of the molecule [14].

Results and Discussion

Optimized geometric and electronic structures, the general energy and electronic energy of molecules (4-methylpentene-1, 4-methylhexen-1, 4,4-dimethylpentene-1) are received by method MNDO and are shown in tables 3.2.5-3.2.8 and on drawings 3.2.4-3.2.6. pKa is certain under the formula pKa=42.11-147.18q_{max}^{H+} [9] (q_{max}^{H+}=+0.05 - the maximal charge on atom of hydrogen, pKa - a universal parameter of acidity) pKa=34,8.

Thus, quantum-chemical calculation of molecules (4-methylpentene-1, 4-methylhexen-1, 4,4-dimethylpentene-1) for the first time it is executed by method MNDO. Optimized geometric and electronic structures of these compounds were received. Acid power was theoretically evaluated. pKa=34,8. We have positioned, that these molecules concern to a class of very weak H-acids (pKa> 14).

Table 3.2.5. Optimized lengths of bonds, valency corners, charges of atoms of the molecule of 4-methylpentene-1

Length of bond	R, A	Valency corners	Grad	Atom	Charge on the atom of molecule
C(1)-C(3)	1,50	C(2)-C(1)-C(3)	127	C(1)	-0.12
C(2)-C(1)	1,34	C(1)-C(3)-C(4)	117	C(2)	-0.05
C(3)-C(4)	1,55	C(3)-C(4)-C(5)	113	C(3)	+0.05
C(4)-C(5)	1,54	C(3)-C(4)-C(6)	113	C(4)	-0.06
C(6)-C(4)	1,54	C(4)-C(5)-H(7)	111	C(5)	+0.04
H(7)-C(5)	1,11	C(4)-C(5)-H(8)	113	C(6)	+0.04
H(8)-C(5)	1,11	C(3)-C(4)-H(9)	105	H(7)	-0.01
H(9)-C(4)	1,12	C(4)-C(6)-H(10)	111	H(8)	0.00
H(10)-C(6)	1,11	C(4)-C(6)-H(11)	113	H(9)	0.01
H(11)-C(6)	1,11	C(1)-C(3)-H(12)	107	H(10)	-0.01
H(12)-C(3)	1,12	C(1)-C(3)-H(13)	110	H(11)	0.00
H(13)-C(3)	1,12	C(2)-C(1)-H(14)	119	H(12)	0.00
H(14)-C(1)	1,10	C(1)-C(2)-H(15)	122	H(13)	0.00
H(15)-C(2)	1,09	C(1)-C(2)-H(16)	124	**H(14)**	**+0.05**
H(16)-C(2)	1,09	C(4)-C(5)-H(17)	111	H(15)	+0.04
H(17)-C(5)	1,11	C(4)-C(6)-H(18)	110	H(16)	+0.04
H(18)-C(6)	1,11			H(17)	-0.01
				H(18)	-0.01

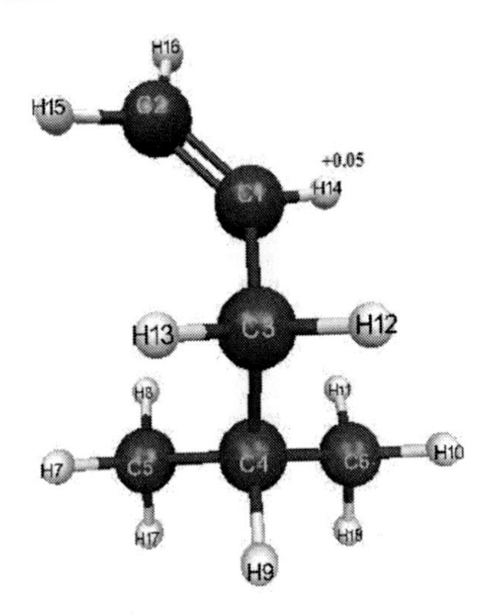

Figure 3.2.4. Geometric and electronic structure of the molecule of 4-methylpentene-1 (E0= -90315 kDg/mol, Eel= 396091 kDg/mol).

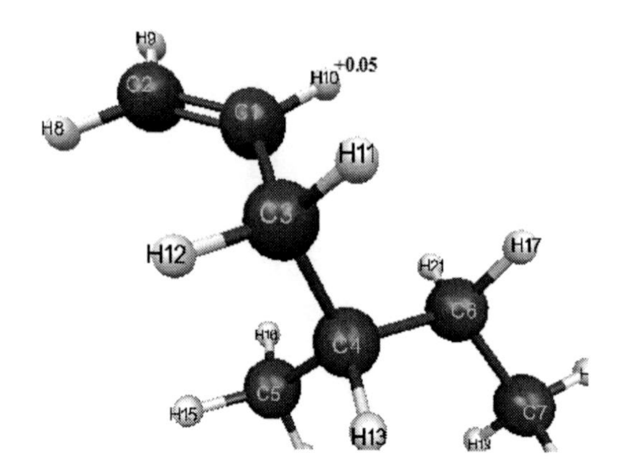

Figure 3.2.5. Geometric and electronic structure of the molecule of 4-methylhexen-1 (E0= -105385 kDg/mol, Eel= -502209 kDg/mol).

Table 3.2.6. Optimized lengths of bonds, valency corners, charges of atoms of the molecule 4-methylhexen-1

Length of bond	R, A	Valency corners	Grad	Atom	Charge on the atom of molecule
C(1)-C(3)	1.50	C(2)-C(1)-C(3)	127	C(1)	-0.11
C(2)-C(1)	1.34	C(1)-C(3)-C(4)	117	C(2)	-0.05
C(3)-C(4)	1.56	C(3)-C(4)-C(5)	113	C(3)	+0.04
C(4)-C(5)	1.54	C(3)-C(4)-C(6)	113	C(4)	-0.05
C(6)-C(4)	1.55	C(4)-C(6)-C(7)	116	C(5)	+0.04
C(7)-C(6)	1.53	C(1)-C(2)-H(8)	122	C(6)	-0.01
H(8)-C(2)	1.09	C(1)-C(2)-H(9)	124	C(7)	+0.03
H(9)-C(2)	1.09	C(2)-C(1)-H(10)	119	H(8)	+0.04
H(10)-C(1)	1.10	C(1)-C(3)-H(11)	107	H(9)	+0.04
H(11)-C(3)	1.12	C(1)-C(3)-H(12)	110	**H(10)**	**+0.05**
H(12)-C(3)	1.12	C(3)-C(4)-H(13)	104	H(11)	0.00
H(13)-C(4)	1.12	C(4)-C(5)-H(14)	111	H(12)	0.00
H(14)-C(5)	1.11	C(4)-C(5)-H(15)	111	H(13)	+0.01
H(15)-C(5)	1.10	C(4)-C(5)-H(16)	113	H(14)	-0.01
H(16)-C(5)	1.11	C(4)-C(6)-H(17)	109	H(15)	-0.01
H(17)-C(6)	1.12	C(6)-C(7)-H(18)	112	H(16)	0.00
H(18)-C(7)	1.11	C(6)-C(7)-H(19)	110	H(17)	0.00
H(19)-C(7)	1.11	C(6)-C(7)-	112	H(18)	0.00
H(20)-C(7)	1.11	H(20)	110	H(19)	-0.01
H(21)-C(6)	1.11	C(4)-C(6)-		H(20)	-0.01
		H(21)		H(21)	+0.01

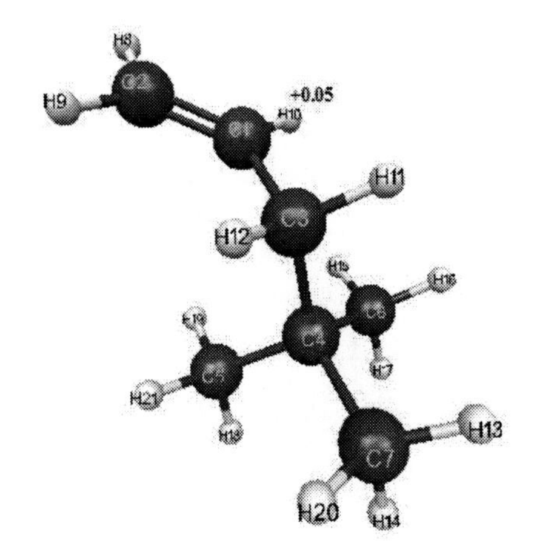

Figure 3.2.6. Geometric and electronic structure of the molecule of 4,4-dimethylpentene-1 (E0= -105358 kDg/mol, Eel= -512326 kDg/mol).

Table 3.2.7. Optimized lengths of bonds, valency corners, charges of atoms of the molecule 4,4-dimethylpentene-1

Length of bond	R, A	Valency corners	Grad	Atom	Charge on the atom of molecule
C(1)-C(3)	1,34	C(2)-C(1)-C(3)	127	C(1)	-0.12
C(2)-C(1)	1,50	C(1)-C(3)-C(4)	117	C(2)	-0.05
C(3)-C(4)	1,57	C(3)-C(4)-C(5)	111	C(3)	+0.05
C(4)-C(5)	1,55	C(3)-C(4)-C(6)	111	C(4)	-0.11
C(6)-C(4)	1,55	C(3)-C(4)-C(7)	108	C(5)	+0.05
C(7)-C(4)	1,56	C(1)-C(2)-H(8)	125	C(6)	+0.05
H(8)-C(2)	1,09	C(1)-C(2)-H(9)	122	C(7)	+0.05
H(9)-C(2)	1,09	C(2)-C(1)-H(10)	119	H(8)	+0.04
H(10)-C(1)	1,10	C(1)-C(3)-H(11)	109	H(9)	+0.04
H(11)-C(3)	1,12	C(1)-C(3)-H(12)	107	H(10)	+0.05
H(12)-C(3)	1,12	C(4)-C(7)-H(13)	112	H(11)	0.00
H(13)-C(7)	1,11	C(4)-C(7)-H(14)	112	H(12)	0.00
H(14)-C(7)	1,11	C(4)-C(6)-H(15)	112	H(13)	-0.01
H(15)-C(6)	1,11	C(4)-C(6)-H(16)	112	H(14)	-0.01
H(16)-C(6)	1,11	C(4)-C(6)-H(17)	111	H(15)	-0.01
H(17)-C(6)	1,11	C(4)-C(5)-H(18)	111	H(16)	0.00
H(18)-C(5)	1,11	C(4)-C(5)-H(19)	112	H(17)	-0.01
H(19)-C(5)	1,11	C(4)-C(7)-H(20)	111	H(18)	-0.01
H(20)-C(7)	1,11	C(4)-C(5)-H(21)	112	H(19)	0.00
H(21)-C(5)	1,11			H(20)	0.00
				H(21)	0.00

Table 3.2.8. The General energy -E_0, the electronic energy of bond (E_{el}), the maximal charge on atom of hydrogen (q_H^{max}), the universal parameter of acidity (pKa) of monomers of cationic polymerization, branched out in β-position in relation to double bond

№	Monomer	-E_0 kDg/mol	-E_{el} kDg/mol	q_{max}^{H+}	pKa
1	4-methylpentene-1	90315	396091	+0,05	34,8
2	4-methylhexen-1	105385	502209	+0,05	34,8
3	4,4-dimethylpentene-1	105358	512326	+0,05	34,8

3.3. Quantum-Chemical Calculation of Linear Olefins of Cationic Polymerization, Branched Out in Γ-Positions in Relation to Double Bond

3.3.1. Calculation by Method AB INITIO

The Aim of this chapter is quantum-chemical calculation of molecules (5-methylhexene-1, 5-methylheptene-1, 6-methylheptene-1) by method AB INITIO in base 6-311G ** with optimization of the geometries by all parameters by standard gradient method in approach the insulated molecule in gas phase and a theoretical estimation of their acid force. The Method is built in in PC GAMESS [12.] Program MacMolPlt was used for visual presentation of the model of the molecule [14].

Results and Discussion

Optimized geometric and electronic structures, the general energy and electronic energy of molecules (5-methylhexene-1, 5-methylheptene-1, 6-methylheptene-1) are received by method AB INITIO in basis 6-311G ** and are shown in tables 3.3.1-3.3.3 and on drawings 3.3.1-3.3.3. pKa is certain under the formula pKa = 49,04 - 134,6q_{max}^{H+} [9]. q_{max}^{H+}= +0,10. pKa=36.

Thus, quantum-chemical calculation of molecules (5-methylhexene-1, 5-methylheptene-1, 6-methylheptene-1) for the first time it is executed by method AB INITIO in base 6-311G** [1]. Optimized geometric and electronic structures of these compounds were received. Acid power was theoretically evaluated. pKa=36. We have positioned, that these molecules concern to a class of very weak H-acids (pKa> 14).

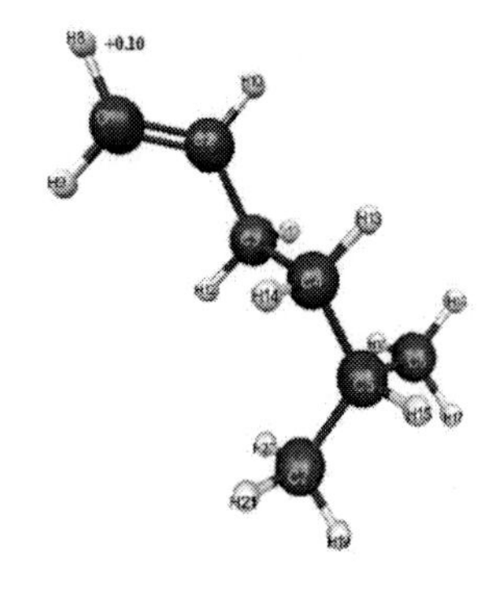

Figure 3.3.1. Geometric and electronic structure of the molecule of 5-methylhexene-1 (E0= -716202 kDg/mol, Eel= -2520743 kDg/mol).

Table 3.3.1. Optimized lengths of bonds, valency corners, charges of atoms of the molecule of 5-methylhexene-1

Length of bond	R, A	Valency corners	Grad	Atom	Charge on the atom of molecule
C(1)-C(2)	1,32	C(1)-C(2)-C(3)	125	C(1)	-0.19
C(2)-C(3)	1,51	C(2)-C(3)-C(4)	112	C(2)	-0.15
C(3)-C(4)	1,54	C(3)-C(4)-C(5)	116	C(3)	-0.15
C(4)-C(5)	1,54	C(4)-C(5)-C(6)	113	C(4)	-0.19
C(5)-C(6)	1,53	C(4)-C(5)-C(7)	113	C(5)	-0.19
C(7)-C(5)	1,53	C(2)-C(1)-H(8)	122	C(6)	-0.22
H(8)-C(1)	1,08	C(2)-C(1)-H(9)	122	C(7)	-0.22
H(9)-C(1)	1,08	C(1)-C(2)-H(10)	119	**H(8)**	**+0.10**
H(10)-C(2)	1,08	C(2)-C(3)-H(11)	109	H(9)	+0.10
H(11)-C(3)	1,09	C(2)-C(3)-H(12)	108	H(10)	+0.09
H(12)-C(3)	1,09	C(3)-C(4)-H(13)	109	H(11)	+0.10
H(13)-C(4)	1,09	C(3)-C(4)-H(14)	108	H(12)	+0.10
H(14)-C(4)	1,09	C(4)-C(5)-H(15)	106	H(13)	+0.09
H(15)-C(5)	1,09	C(5)-C(6)-H(16)	111	H(14)	+0.10
H(16)-C(6)	1,09	C(5)-C(6)-H(17)	111	H(15)	+0.10
H(17)-C(6)	1,09	C(5)-C(6)-H(18)	112	H(16)	+0.09
H(18)-C(6)	1,09	C(5)-C(7)-H(19)	111	H(17)	+0.09
H(19)-C(5)	1,09	C(5)-C(7)-H(20)	112	H(18)	+0.08
H(20)-C(5)	1,09	C(5)-C(7)-H(21)	111	H(19)	+0.09
H(21)-C(5)	1,09			H(20)	+0.09
				H(21)	+0.09

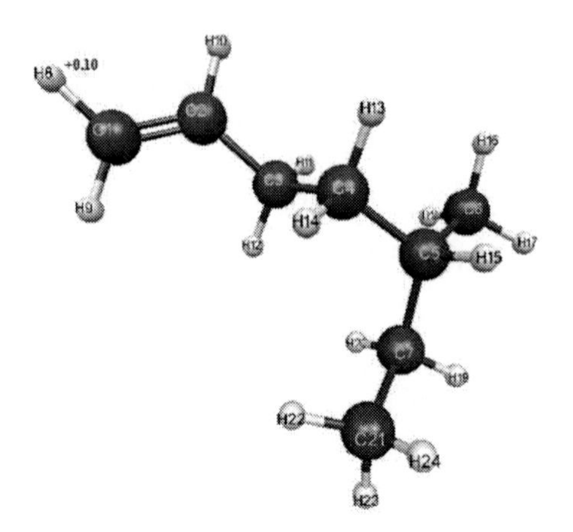

Figure 3.3.2. Geometric and electronic structure of the molecule of 5-methylheptene-1 (E0= -818520 kDg/mol, Eel= -3050812 kDg/mol).

Table 3.3.2. Optimized lengths of bonds, valency corners, charges of atoms of the molecule of 5-methylheptene-1

Length of bond	R, A	Valency corners	Grad	Atom	Charge on the atom of molecule
C(1)-C(2)	1,32	C(1)-C(2)-C(3)	125	C(1)	-0.20
C(2)-C(3)	1,51	C(2)-C(3)-C(4)	112	C(2)	-0.15
C(3)-C(4)	1,54	C(3)-C(4)-C(5)	116	C(3)	-0.15
C(4)-C(5)	1,54	C(4)-C(5)-C(6)	112	C(4)	-0.19
C(5)-C(6)	1,53	C(4)-C(5)-C(7)	114	C(5)	-0.19
C(7)-C(5)	1,54	C(2)-C(1)-H(8)	122	C(6)	-0.22
H(8)-C(1)	1,08	C(2)-C(1)-H(9)	122	C(7)	-0.19
H(9)-C(1)	1,08	C(1)-C(2)-H(10)	119	**H(8)**	**+0.10**
H(10)-C(2)	1,08	C(2)-C(3)-H(11)	109	H(9)	+0.10
H(11)-C(3)	1,09	C(2)-C(3)-H(12)	109	H(10)	+0.10
H(12)-C(3)	1,09	C(3)-C(4)-H(13)	109	H(11)	+0.10
H(13)-C(4)	1,09	C(3)-C(4)-H(14)	109	H(12)	+0.10
H(14)-C(4)	1,09	C(4)-C(5)-H(15)	106	H(13)	+0.10
H(15)-C(5)	1,09	C(5)-C(6)-H(16)	111	H(14)	+0.10
H(16)-C(6)	1,09	C(5)-C(6)-H(17)	111	H(15)	+0.10
H(17)-C(6)	1,09	C(5)-C(6)-H(18)	112	H(16)	+0.09
H(18)-C(6)	1,09	C(5)-C(7)-H(19)	108	H(17)	+0.09
H(19)-C(7)	1,09	C(5)-C(7)-H(20)	110	H(18)	+0.09
H(20)-C(7)	1,09	C(5)-C(7)-C(21)	115	H(19)	+0.09
C(21)-C(7)	1,53	C(7)-C(21)-H(22)	112	H(20)	+0.09
H(22)-C(21)	1,09	C(7)-C(21)-H(23)	111	C(21)	-0.23
H(23)-C(21)	1,09	C(7)-C(21)-H(24)	111	H(22)	+0.09
H(24)-C(21)	1,09			H(23)	+0.09
				H(24)	+0.09

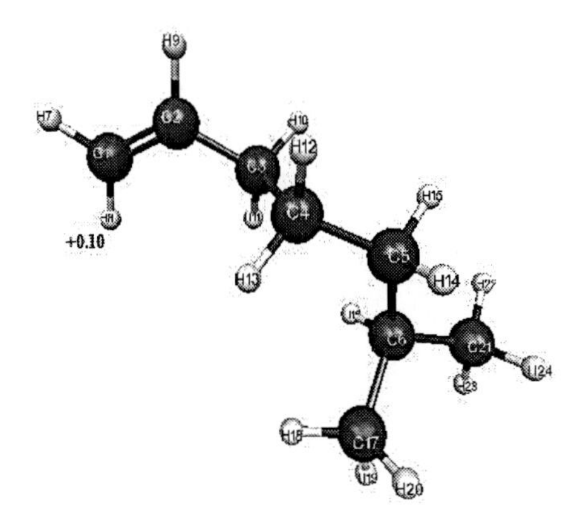

Figure 3.3.3. Geometric and electronic structure of the molecule of 6-methylheptene-1 (E0= -818521 kDg/mol, Eel= -3019388 kDg/mol).

Table 3.3.3. Optimized lengths of bonds, valency corners, charges of atoms of the molecule of 6-methylheptene-1

Length of bond	R, A	Valency corners	Grad	Atom	Charge on the atom of molecule
C(1)-C(2)	1,32	C(1)-C(2)-C(3)	125	C(1)	-0.19
C(2)-C(3)	1,5	C(2)-C(3)-C(4)	112	C(2)	-0.15
C(3)-C(4)	1,53	C(3)-C(4)-C(5)	115	C(3)	-0.15
C(4)-C(5)	1,53	C(4)-C(5)-C(6)	119	C(4)	-0.19
C(5)-C(6)	1,54	C(2)-C(1)-H(7)	122	C(5)	-0.17
H(7)-C(6)	1,08	C(2)-C(1)-H(8)	122	C(6)	-0.21
H(8)-C(1)	1,08	C(1)-C(2)-H(9)	119	C(7)	+0.10
H(9)-C(2)	1,08	C(2)-C(3)-H(10)	109	**H(8)**	**+0.10**
H(10)-C(3)	1,09	C(2)-C(3)-H(11)	109	H(9)	+0.09
H(11)-C(3)	1,09	C(3)-C(4)-H(12)	109	H(10)	+0.10
H(12)-C(4)	1,09	C(3)-C(4)-H(13)	109	H(11)	+0.10
H(13)-C(4)	1,09	C(4)-C(5)-H(14)	108	H(12)	+0.09
H(14)-C(5)	1,09	C(4)-C(5)-H(15)	108	H(13)	+0.10
H(15)-C(5)	1,09	C(5)-C(6)-H(16)	108	H(14)	+0.09
H(16)-C(6)	1,09	C(5)-C(6)-C(17)	113	H(15)	+0.09
C(17)-C(6)	1,53	C(6)-C(17)-H(18)	112	H(16)	+0.10
H(18)-C(17)	1,08	C(6)-C(17)-H(19)	111	H(17)	-0.22
H(19)-C(17)	1,09	C(6)-C(17)-H(20)	111	H(18)	+0.10
H(20)-C(17)	1,09	C(5)-C(6)-C(21)	110	H(19)	+0.09
C(21)-C(6)	1,53	C(6)-C(21)-H(22)	111	H(20)	+0.09
H(22)-C(21)	1,09	C(6)-C(21)-H(23)	111	C(21)	-0.22
H(23)-C(21)	1,09	C(6)-C(21)-H(24)	111	H(22)	+0.09
H(24)-C(21)	1,09			H(23)	+0.09
				H(24)	+0.08

Table 3.3.4. The General energy -E_0, the electronic energy of bond (E_{el}), the maximal charge on atom of hydrogen (q_H^{max}), the universal parameter of acidity (pKa) of monomers of cationic polymerization, branched out in γ-position in relation to double bond

№	Monomer	-E_0 kDg/mol	-E_{el} kDg/mol	q_{max}^{H+}	pKa
1	5-methylhexene-1	716202	2520743	+0,10	36
2	5-methylheptene-1	818520	3050812	+0,10	36
3	6-methylheptene-1	818521	3019388	+0,10	36

3.3.2. Calculation by Method MNDO

The Aim of this chapter is quantum-chemical calculation of molecules of olefins of cationic polymerization, branched out in γ-position in relation to double bond (5-methylhexene-1, 5-methylheptene-1, 6-methylheptene-1) by method MNDO with optimization of the geometries by all parameters by standard gradient method in approach the insulated molecule in gas phase and a theoretical estimation of their acid force. The Method is built in in PC GAMESS [12.] Program MacMolPlt was used for visual presentation of the model of the molecule [14].

RESULTS AND DISCUSSION

Optimized geometric and electronic structures, the general energy and electronic energy of molecules 5-methylhexene-1, 5-methylheptene-1, 6-methylheptene-1 are received by method MNDO and are shown in tables 3.3.5-3.3.8 and on drawings 3.3.4-3.3.6. pKa is certain under the formula $pKa=42.11-147.18q_{max}^{H+}$ [9] ($q_{max}^{H+}=+0.05$ - the maximal charge on atom of hydrogen, pKa - a universal parameter of acidity) pKa=35.

Thus, quantum-chemical calculation of molecules 5-methylhexene-1, 5-methylheptene-1, 6-methylheptene-1 for the first time it is executed by method MNDO. Optimized geometric and electronic structures of these compounds were received. Acid power was theoretically evaluated. pKa=35. We have positioned, that these molecules concern to a class of very weak H-acids (pKa> 14).

Table 3.3.5. Optimized lengths of bonds, valency corners, charges of atoms of the molecule of 5-methylhexene-1

Length of bond	R, A	Valency corners	Grad	Atom	Chargeon the atom of molecule
C(1)-C(2)	1,34	C(1)-C(2)-C(3)	127	C(1)	-0.05
C(2)-C(3)	1,51	C(2)-C(3)-C(4)	113	C(2)	-0.12
C(3)-C(4)	1,54	C(3)-C(4)-C(5)	118	C(3)	+0.03
C(4)-C(5)	1,55	C(4)-C(5)-C(6)	114	C(4)	+0.01
C(5)-C(6)	1,54	C(4)-C(5)-C(7)	114	C(5)	-0.07
C(7)-C(5)	1,54	C(2)-C(1)-H(8)	122	C(6)	+0.04
H(8)-C(1)	1,09	C(2)-C(1)-H(9)	124	C(7)	+0.04
H(9)-C(1)	1,09	C(1)-C(2)-H(10)	119	H(8)	+0.04
H(10)-C(2)	1,10	C(2)-C(3)-H(11)	108	H(9)	+0.04
H(11)-C(3)	1,11	C(2)-C(3)-H(12)	110	**H(10)**	**+0.05**
H(12)-C(3)	1,11	C(3)-C(4)-H(13)	109	H(11)	+0.01
H(13)-C(4)	1,12	C(3)-C(4)-H(14)	108	H(12)	0.00
H(14)-C(4)	1,12	C(4)-C(5)-H(15)	104	H(13)	0.00
H(15)-C(5)	1,12	C(5)-C(6)-H(16)	111	H(14)	0.00
H(16)-C(6)	1,09	C(5)-C(6)-H(17)	111	H(15)	+0.01
H(17)-C(6)	1,10	C(5)-C(6)-H(18)	113	H(16)	-0.01
H(18)-C(6)	1,11	C(5)-C(7)-H(19)	111	H(17)	-0.01
H(19)-C(5)	1,11	C(5)-C(7)-H(20)	113	H(18)	-0.01
H(20)-C(5)	1,11	C(5)-C(7)-H(21)	111	H(19)	-0.01
H(21)-C(5)	1,11			H(20)	-0.01
				H(21)	-0.01

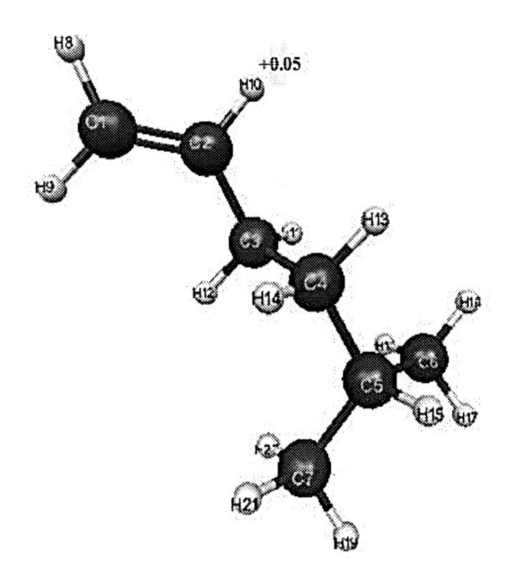

Figure 3.3.4. Geometric and electronic structure of the molecule of 5-methylhexene-1 (E0= -105377 kDg/mol, Eel= -496918 kDg/mol).

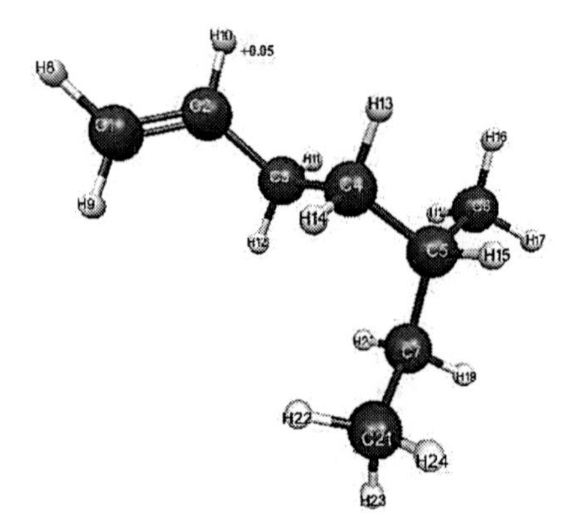

Figure 3.3.5. Geometric and electronic structure of the molecule of 5-methylheptene-1 (E0= -120428 kDg/mol, Eel= -613019 kDg/mol).

Table 3.3.6. Optimized lengths of bonds, valency corners, charges of atoms of the molecule of 5-methylheptene-1

Length of bond	R, A	Valency corners	Grad	Atom	Charge on the atom of molecule
C(1)-C(2)	1,34	C(1)-C(2)-C(3)	126	C(1)	-0.05
C(2)-C(3)	1,51	C(2)-C(3)-C(4)	113	C(2)	-0.12
C(3)-C(4)	1,54	C(3)-C(4)-C(5)	119	C(3)	+0.03
C(4)-C(5)	1,55	C(4)-C(5)-C(6)	113	C(4)	+0.01
C(5)-C(6)	1,54	C(4)-C(5)-C(7)	116	C(5)	-0.05
C(7)-C(5)	1,55	C(2)-C(1)-H(8)	122	C(6)	+0.04
H(8)-C(1)	1,09	C(2)-C(1)-H(9)	124	C(7)	-0.01
H(9)-C(1)	1,09	C(1)-C(2)-H(10)	119	H(8)	+0.04
H(10)-C(2)	1,10	C(2)-C(3)-H(11)	108	H(9)	+0.04
H(11)-C(3)	1,11	C(2)-C(3)-H(12)	110	**H(10)**	**+0.05**
H(12)-C(3)	1,11	C(3)-C(4)-H(13)	108	H(11)	+0.01
H(13)-C(4)	1,12	C(3)-C(4)-H(14)	108	H(12)	+0.01
H(14)-C(4)	1,11	C(4)-C(5)-H(15)	104	H(13)	0.00
H(15)-C(5)	1,12	C(5)-C(6)-H(16)	111	H(14)	0.00
H(16)-C(6)	1,11	C(5)-C(6)-H(17)	111	H(15)	0.00
H(17)-C(6)	1,11	C(5)-C(6)-H(18)	113	H(16)	-0.01
H(18)-C(6)	1,11	C(5)-C(7)-H(19)	118	H(17)	-0.01
H(19)-C(7)	1,12	C(5)-C(7)-H(20)	110	H(18)	0.00
H(20)-C(7)	1,11	C(5)-C(7)-C(21)	117	H(19)	0.00
C(21)-C(7)	1,53	C(7)-C(21)-H(22)	112	H(20)	+0.01
H(22)-C(21)	1,11	C(7)-C(21)-H(23)	110	C(21)	+0.03
H(23)-C(21)	1,11	C(7)-C(21)-H(24)	112	H(22)	-0.01
H(24)-C(21)	1,11			H(23)	-0.01
				H(24)	0.00

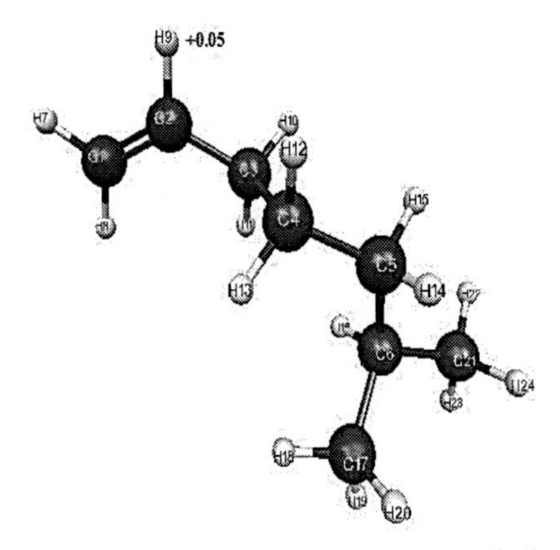

Figure 3.3.6. Geometric and electronic structure of the molecule of 6-methylheptene-1 (E0= -120454 kDg/mol, Eel= -605340 kDg/mol).

Table 3.3.7. Optimized lengths of bonds, valency corners, charges of atoms of the molecule of 6-methylheptene-1

Length of bond	R, A	Valency corners	Grad	Atom	Charge on the atom of molecule
C(1)-C(2)	1,34	C(1)-C(2)-C(3)	127	C(1)	-0.05
C(2)-C(3)	1,51	C(2)-C(3)-C(4)	113	C(2)	-0.12
C(3)-C(4)	1,54	C(3)-C(4)-C(5)	116	C(3)	+0.03
C(4)-C(5)	1,54	C(4)-C(5)-C(6)	118	C(4)	-0.01
C(5)-C(6)	1,55	C(2)-C(1)-H(7)	122	C(5)	+0.01
H(7)-C(6)	1,09	C(2)-C(1)-H(8)	124	C(6)	-0.07
H(8)-C(1)	1,09	C(1)-C(2)-H(9)	119	C(7)	+0.04
H(9)-C(2)	1,10	C(2)-C(3)-H(10)	108	H(8)	+0.04
H(10)-C(3)	1,12	C(2)-C(3)-H(11)	110	**H(9)**	**+0.05**
H(11)-C(3)	1,11	C(3)-C(4)-H(12)	108	H(10)	0.00
H(12)-C(4)	1,12	C(3)-C(4)-H(13)	109	H(11)	0.00
H(13)-C(4)	1,12	C(4)-C(5)-H(14)	107	H(12)	0.00
H(14)-C(5)	1,12	C(4)-C(5)-H(15)	108	H(13)	+0.01
H(15)-C(5)	1,12	C(5)-C(6)-H(16)	108	H(14)	0.00
H(16)-C(6)	1,12	C(5)-C(6)-C(17)	114	H(15)	0.00
C(17)-C(6)	1,54	C(6)-C(17)-H(18)	112	H(16)	+0.01
H(18)-C(17)	1,11	C(6)-C(17)-H(19)	111	H(17)	+0.04
H(19)-C(17)	1,11	C(6)-C(17)-H(20)	112	H(18)	-0.01
H(20)-C(17)	1,11	C(5)-C(6)-C(21)	111	H(19)	-0.01
C(21)-C(6)	1,54	C(6)-C(21)-H(22)	111	H(20)	0.00
H(22)-C(21)	1,11	C(6)-C(21)-H(23)	111	C(21)	+0.04
H(23)-C(21)	1,11	C(6)-C(21)-H(24)	112	H(22)	0.00
H(24)-C(21)	1,11			H(23)	0.00
				H(24)	0.00

Table 3.2.8. The General energy -E_0, the electronic energy of bond (E_{el}), the maximal charge on atom of hydrogen (q_H^{max}), the universal parameter of acidity (pKa) of monomers of cationic polymerization, branched out in γ-position in relation to double bond

№	Monomer	-E_0 kDg/mol	-E_{el} kDg/mol	q_{max}^{H+}	pKa
1	5-methylhexene-1	105377	496918	+0,05	35
2	5-methylheptene-1	120428	613019	+0,05	35
3	6-methylheptene-1	120454	605340	+0,05	35

CONCLUSION

Quantum-chemical calculations of linear monomer of cationic polymerization, branched out in α-, β-, γ-position in relation to double bond for the first time are executed in the present work. Classical methods of quantum chemistry MNDO (one of the fastest now) and AB INITIO (the most exact and the best for today) were used for calculation with optimization of geometry on all parameters by standard gradient method. The optimized geometrical and electronic structure of these compounds is received. Acid force of these monomers is theorized. We have positioned, that all of them possess very low acid force (values pKa of all studied monomers are in range from 24 up to 38) and concern to a class of very weak H-acids (as pKa more than 14).

The analysis of the received values of acid force pKa of linear monomers of cationic polymerization will allow to position its influence on mechanisms of initiation, growth and breakage of a circuit in catalytic processes of reception of polymers, or absence of this influence.

REFERENCES

[1] Babkin V. A. Quantum chemical calculation of cationic polymerization of olefins / V. A. Babkin, K. S. Minsker, G. E. Zaikov. – New–York, 1997. – P. 138.

[2] Babkin V. A. Quantum chemical aspects of cationic polymerization of olefins / V. A. Babkin, K. S. Minsker, G. E. Zaikov. – Ufa, 1996. – P. 181.

[3] Sangalov U. A. Polymers and copolymers of isobutylene / U. A. Sangalov, K. S. Minsker. – Ufa, 2001. – P. 381.

[4] Pople I. A. In approximate Molecular Orbit Theory / I. A. Pople, D. N. Beveridze. – Y. Mc. Graw - Hill, 1970. – P. 214.

[5] Zhidomirov T. M. The Applied quantum chemistry / T. M. Zhidomirov, A. A. Bagaturianc, I. A. Abronin. – M. : Chemistry, 1979. – P. 255.

[6] Martin J. M. L. Critical Comparison of MINDO/3, MNDO, AM1, and PM3 for a Model Problems Carbon Clusters C_2-C_{10}. An Ad Hoc Reparametrization of MNDO Well Suited for Accurate Prediction of Their Spectroscopic Constants / J. M. L. Martin, J. P. Francois, R. A. Gijbels, // J. Comp. Chem. – 1991. – Vol. 12. – No. 1. – P. 52–70.

[7] Klark T. The Computer chemistry / T. Klark. – M.: World, 1990. – P. 383.

[8] Stepanov N. F. Quantum mechanics of molecules and quantum chemistry / N. F. Stepanov, V. I. Pupushev. – 1991.

[9] Babkin V. A. Connection of the universal acidity index of H-acids with the charge on hydrogen atom (AB INITIO METHOD).Oxidation communication / V. A. Babkin, R. G. Fedunov, K. S. Minsker [and another]. – 2002. – №1, 25. – P. 21 - 47.

[10] Babkin V. A. Computer nanotechnologys of applied quantum chemistry / Babkin V. A., R. G. Fedunov. – Volgograd : VolgSABU, 2008. – P. 135.

[11] Kennedi G. The Cationic polymerization of olefins / G. Kennedi. – M., 1978. – P. 431.

[12] Shmidt M.W. J. Comput. Chem. / M. W. Shmidt, M. S. Gordon [and another]. – 1993. – 14. – P. 1347-1363.

[13] Meshkov A. N. The Estimation of acid force of linear olefins and nonenveloping diolefins / A. N. Meshkov, S. N. Slushkin, D. A. Skvorcov // Theoretical and applied aspects of modern natural history : collection of scientific articles – Volgograd : SD VolgSABU, 2005. – P. 42-53.

[14] Bode B. M. J. Mol. Graphics Mod / B. M. Bode, M. S. Gordon. – 1998. – 16. – P.133-138.

INDEX

A

B

C

D

E

F

G

H

I

L

M

N

O

P